How History Gets Things Wrong

How History Gets Things Wrong

The Neuroscience of Our Addiction to Stories

Alex Rosenberg

The MIT Press
Cambridge, Massachusetts
London, England

This book was set in Stone Serif by Westchester Publishing Services. Printed and bound in the United States of America.

Library of Congress Cataloging-in-Publication Data is available.

ISBN: 978-0-262-03857-7

10 9 8 7 6 5 4 3 2 1

Contents

1 Besotted by Stories 1

2 How Many Times Can the German Army
Play the Same Trick? 15

3 Why Ever Did Hitler Declare War on the United States?
That's Easy to Explain, Too Easy 31

4 Is the Theory of Mind Wired In? 49

5 The Natural History of Historians 65

6 What Exactly Was the Kaiser Thinking? 95

7 Can Neuroscience Tell Us What Talleyrand Meant? 111

8 Talleyrand's Betrayal: The Inside Story 141

9 *Jeopardy!* "Question": "It Shows the Theory of Mind
to Be Completely Wrong" 163

10 The Future of an Illusion 185

11 Henry Kissinger Mind Reads His Way through
the Congress of Vienna 209

12 *Guns, Germs, Steel*—and All That 219

13 *The Gulag Archipelago* and the Uses of History 241

The Back(non)story 251
References 261
Index 273

1 Besotted by Stories

It's almost universally accepted that learning the history of something—the true story of how it came about—is one way to understand it. It's almost as widely accepted that learning its history is sometimes the best way to understand something. Indeed, in many cases, it's supposed that the *only* way to understand some things is by learning their history.

How History Gets Things Wrong explains why all three of these suppositions are wrong. Cognitive science, evolutionary anthropology, and, most of all, neuroscience are in the process of showing us at least three things about history: (1) our attachment to history as a vehicle for understanding has a long evolutionary pedigree and a genetic basis; (2) exactly what it is about the human brain that makes almost all the explanations history has ever offered us wrong; and (3) how our evolution shaped a useful tool for survival into a defective theory of human nature.

Many readers may find the first of these assertions easy to accept. Our recourse to history—true stories—as a means of understanding is proverbially "second nature." If science can show it's literally "first nature," bred in the bone, a part of what makes us tick, somehow genetically hardwired, it may help us understand features of human life and culture that are ancient, ubiquitous, and fixed beyond change. But the next two assertions will strike most readers as literally incredible. How could all the explanations history offers be wrong, and how could evolution by itself have saddled us with any particular theory, let alone a theory of human nature that is completely wrong?

The three assertions—that our confidence in history, our taste, our need for it, indeed, our love of history is almost completely hardwired, that history is all wrong, and that its wrongness is the result of the later evolution of what was originally hardwired—are pretty much a package deal. The second and third hard-to-accept assertions build on the first one, and they do so in ways that make them hard to reject. If cognitive psychology, evolutionary anthropology, and, most of all, neuroscience between them explain why we are so attached to history as a way of understanding, they also undermine history's claim to provide real understanding of the past, the present and the future.

Just to be clear, historians are perfectly capable of establishing actual, accurate, true chronologies and other facts about what happened in the past. They aren't wrong about feudalism coming before the Reformation or whether Italy and Japan were on the Allies' side in World War One. Moreover, historians working in archives, for example, retrieve documentary evidence for important events in human history that have disappeared from collective memory or were never even noticed. More important, many written histories, especially those produced in the academic departments of universities, are more than just accurate chronicles. The approaches to the past that many professors of history employ can provide powerful new and better explanations of well-known historical events and processes, often by identifying causes previously unknown or ignored (as we will see). Academic history is more than, and usually different from, true stories.

But academic history isn't the history that we consume to explain individual human actions and the lives they constitute, or to understand famous creative, political, public, and scientific achievements, fateful choices and their all too often tragic consequences. That's because nowadays academic history is rarely narrative. The history that professors write these days has been deeply influenced by the sciences—social, behavioral, even natural—and it rarely seeks to explain individual achievements or even lives, singly or taken together. Academic history often makes use of stories—records, letters, diaries, chronicles that people write down—as evidence for its explanations. But it is not much given to explaining by telling these (true) stories.

The history that concerns us here explains the past and the present by narrative: telling stories—true ones, of course; that's what makes them history, not fiction. Narrative history is not just an almanac or a chronology of what happened in the past. It is explanation of what happened in terms of the motives and the perspectives of the human agents whose choices, decisions, and actions made those events happen.

And that history, the kind most readers of nonfiction consume, is almost always wrong. What narrative history gets wrong are its *explanations* of what happened. And the same goes for biography—the history of one person over a lifetime. Biographers can get all the facts from birth to death right. What they inevitably get wrong is why their subjects did what they accurately report them as having done.

Just how narrative history gets everything else wrong is, of course, the subject of this book. It all starts with the fact that most history is narrative, narrative is stories, and stories are chronologies stitched together into plots we understand better than anything else, or at least we think we do. The same science that reveals why we view the world through the lens of narrative also shows that the lens not only distorts what we see but is the source of illusions we can neither shake nor even correct for most of the time. As we'll see, however, all narratives are wrong—wrong in the same way and for the same reason.

Uncovering what bedevils all narrative and, through it, all narrative history, is important, perhaps crucial to the future of humanity. How beneficial the impact of that uncovering will be is hard to say, for reasons that will become clear. But it is easy to identify, as indeed we will, the vast harms that have resulted from the hegemony of narrative history in human affairs.

It's crucial to disabuse ourselves of the myth that history confers real understanding that can shape or otherwise help us cope with the future. Almost everyone thinks history is a route to knowledge, sometimes one among many, sometimes the best route, sometimes the only route to it. And once people think they know something, they act on that knowledge. If they're wrong about what they know, the results can be frustration, disappointment, or worse, all the way from harm to themselves up to catastrophes for humanity.

The claim that historical narrative is indispensable to understanding is made casually and daily. Pick up any issue of the *New York Times Book Review*, or the *London* or *New York Review of Books*, or the *Times Literary Supplement*, and you'll find a reviewer lauding a history or a biography as indispensable for understanding some perfectly nonhistorical subject. It's not just historians who say so; often it's the experts on that subject who do. For example, suppose you want to understand the policy of "quantitative easing" employed by the Federal Reserve System of the United States in the aftermath of the financial crisis of 2007–2008. Well, then, you need to read

Roger Lowenstein's history of the Fed's creation 100 years ago, *America's Bank: The Epic Struggle to Create the Federal Reserve* (Lowenstein, 2015). How do we know? Just take the word of former U.S. Secretary of the Treasury Robert Rubin: "It should be *required* reading for anyone who is engaged in, or interested in, the actions of the *modern* Fed" (Rubin, 2015). Pardonable exaggeration?

Are you interested in what made Steve Jobs tick? Then surely you have to get Walter Isaacson's biography (Isaacson, 2015). Or maybe you should also read *Becoming Steve Jobs* by Brent Schlender and Rick Tetzeli (Schlender and Tetzeli, 2015). The truth about what Jobs was thinking in his final months and days must lie somewhere between these two histories of this one man.

Even where historical narrative doesn't seem indispensable to understanding something, it is widely believed to be the best way. Nothing illustrates this belief more clearly than the penchant of science writers for historical narrative. Science is not stories; it's theories, laws, models, findings, observations, experiments. Yet almost the only way writers communicate science to the general public is through the narrative history of breakthroughs or the biographies of the scientists who achieved them. You want to know about the theory of plate tectonics and continental drift, read Simon Winchester's wonderful history of the subject, *Krakatoa* (Winchester, 2003). Even if you don't want to know about the subject, Winchester will make you interested in it. Space-time physics? Well, then the book for you is Stephen Hawking's *A Brief History of Time* (Hawking, 1988), which explains cosmology via a brief history of physics going back to Aristotle. In fact, the number-one best seller in science books is Bill Bryson's *A Short History of Nearly Everything* (Bryson, 2003). There's a good reason writers communicate science in stories. It's not just that most people prefer stories. It's also because most nonscientists find it extremely hard, if not impossible, to acquire scientific information in any other way. The trouble is, after they've read these science books, people generally remember the stories, but they can't recall the science.

It's widely held that history is not just necessary for understanding something, and not just the best way to acquire this understanding (or delude yourself into thinking you have), but that it's *all* you need to understand something in at least one domain—history itself.

It's slightly inconvenient that in English (and other languages) the word "history" is ambiguous: it is used to describe both what happened in the past and the quite different study of what happened in the past. German has two different terms: there's what happened, the events themselves, "historische Ereignisse," and then there is their history, "Geschichte." Throughout this book, we'll have to live with the ambiguity. But with the distinction in mind, it seems obvious that the only way to learn about history, the past, is history, the fruit of the study of the past, and that all you need to understand the past is the right history of it. Thus there are still those who hold that all you need to do to understand the decline and fall of the Roman Empire is to read Edward Gibbon's glorious narrative *The History of the Decline and Fall of the Roman Empire* (Gibbon, 1776–1789).

In the chapters to follow, we'll see how all three of these suppositions about history are wrong: how it's not enough to understand anything, or even just indispensable or necessary for understanding it, in fact how it's not even a good way, let alone the best or only way, to understand anything.

If—and it's a big "if"—narrative history gets almost everything wrong, why does it matter? Because the narratives that the field of history has provided have been harmful to the health, well-being, and the very lives of most people down through the chain of historical events. Stories historians tell are deeply implicated in more misery and death than probably any other aspect of human culture. And, as we'll see, it's the nature of the most compelling stories they tell that's responsible for the trail of tears, pain, suffering, carnage, and sometimes extermination that make up most of human history.

Many disputes between peoples are intractable owing to the histories each side tells itself. Perhaps the most obvious current example of how history hardens two parties in conflict is provided by Israel and Palestine. Each has a narrative of dispossession, one two thousand years old, the other a century old, both of which drive emotions so strongly that neither side can put the histories aside and find a *modus vivendi* with the other. If only the Israelis and Palestinians accepted that stories fan emotional flames rather than confer understanding, they might cease to grant them authority over how the future should be arranged.

For a more egregious example, you have only to think back to the story of the "Stab in the back" historians composed to explain the defeat of Germany in World War One—the "Great War"—and that Adolf Hitler exploited to inflame the German people against Socialists, democrats, and, most of all, Jews. The wars of nationalism, religion, imperialism, colonialism—and anticolonialism, for that matter—begin and persist because of grievances often fueled by historical narratives.

Why do I say that most historical narratives are harmful, damaging to people, instead of saying that some are also uplifting and inspiring? Have I made a count, totting up numbers and weighing the baleful effects of some against the benevolent effects of others? It's not necessary. As you'll see, evolutionary anthropologists know enough about human cultural evolution to be confident that most histories have motivated and continue to motivate people and peoples to take from—or refuse to share with—others. The xenophobia, racism, and patriarchy that ruled long before the advent of the nation-state were already clothed in histories of who did what to whom. The nation-state, when it arrived, was just a more efficient means to raise the death toll of narratives. The Old Testament is only the best known of these vehicles of in-group bonding and out-group enmity. As you'll see, stories emerged in human prehistory as practices that were able to move humans from the bottom of the food chain on the African savanna to the top in a matter of a thousand centuries or so. These cultural practices were selected for owing to their effectiveness, first, in killing large animals and, then, in killing—or, even worse, enslaving—other humans.

There is a reason why stories and histories of war and killing have been more popular than the lives of saints or artists since Homer. The reason is not, however, given in a history. It is provided by science, in this case, social psychology, as we'll see.

If we humans are ever to move beyond our internecine histories, we will have to put historical "understanding" behind us. We will have to recognize that even the best histories we can contrive are mostly wrong or, when right, are right by accident, that they fail to identify the real causal forces that drive events, that they obstruct efforts to really understand our past, and that they serve as harmful tools of the worse angels of our nature.

In this book, you'll see why we love history. Preadapted to it, we took the love of history in with our mothers' milk. More like heroin than milk, this love is an addiction to history, however, not a mere taste for it. Even when we come to recognize its harmful effects, we continue to crave the sensation it produces. And we'll never be comfortable with the only medication that can block its harmful effects—science. Like the recovering alcoholic, once we recognize the disease, we'll also realize that we have to struggle every day not to succumb to its temptations, its all too meretricious allure, if we're ever to really understand anything at all.

Historians will of course be outraged by these assertions, and confident about how to refute them. To begin with, most contemporary academic historians will deny any interest in advancing explanatory narratives—mere "stories," even true ones. And they're right to do so: indeed, what most faculty members produce in the history departments of the world's universities is not the target of this book. But the celebrated popular historians whose explanations turn out to be mainly wrong will protest just as vigorously. They'll present story after story that vindicates their sort of historical understanding, showing that it is not only indispensable, but also the most, indeed the only, reliable guide to what happened in the past and to what will happen in the future. Chapter 2 will give these historians a run for their money, exploring their track record under the most favorable circumstances, where they were paid to provide exactly the sort of historical knowledge that they claim we need to understand our past and foresee our future.

Meanwhile, there are several things we need to consider that should make us skeptical about narrative history as a path to understanding. For one thing, when it comes to physics, geology, and the other natural sciences, the specialists don't care about history much at all. Read the textbooks, scientific journals, attend the seminars and colloquia where they present their results to one another. The histories of their disciplines—how they got to where they are today, don't come into it. Facts, data, evidence, observations are all important, and though many are about past events, recent or distant, all they do is provide evidence for scientific results, findings, models, or theories. Scientists never confuse science with the narrative histories of science, still less with the biographies of scientists.

Why is it that learning the history of a scientific breakthrough is not a way to acquire understanding in the sciences? And why is it that when scientists and science writers seek to communicate science to nonscientists, narrative history is pretty much the only way they try to do so? These are two matters we'll address in this book.

A second thing we need to consider about history that should make us skeptical is the unending disagreement among historians over the same events. Gibbon was hardly the last word on the decline and fall of the Roman Empire. Historians have been arguing inconclusively about this matter since well before the sixth and last volume of Gibbon's history was published in 1789. And the arguments about whether Gibbon got it right—whether he correctly identified the cause of Rome's decline—have not turned on the discovery of new evidence unavailable to him. Narrative historians are forever rewriting the past, disputing one another's causal claims. And there is no reason to think they will ever cease to do so, even for events as long past as the fall of the Roman Empire. Two centuries after Gibbon published the first volume of his history, Mary Beard published her distinctly different account, *SPQR: A History of Ancient Rome* (Beard, 2015), which rose to the top of many best-seller lists soon after. Thus, even after all that time, there's still no agreement on why the Roman Empire fell.

Biography is as much subject to revisionism as history is. There are good reasons to read Walter Isaacson's biography of Steve Jobs. It's entertaining, amusing, even inspiring. But, within weeks of its publication, other books appeared disputing Isaacson's understanding of Jobs. Such disputes will continue until the public is no longer fascinated by their subjects. And there are many figures in whom interest never ends. Indeed, Lincoln has been the subject of some 40,000 books. But even a correct recitation of the facts about someone that links those facts into a narrative of the person's life doesn't seem to settle any matter about what made that someone tick. New archival materials—letters, diaries, eyewitness testimony—may add to the record of events and may even discredit other materials. But, of course, that's not why George Washington or Winston Churchill gets a new biography every generation (or sooner).

Historical revisionism could be evidence supporting narrative history's claim to provide real understanding, but only if, like successive scientific explanations, historical explanations converged. Scientific theories may start off being very wrong, but, over time, at least the ones that survive testing get better, as scientists home in on an ever-smaller number of better explanations of what their predecessors set out to explain. Two important reasons to think that the explanations that survive the winnowing process are better than the ones that don't are their predictive success and their technological applications. By contrast, historians' successive explanations for the same historical events—historical revisionism—don't show the same kind of convergence. Instead, these explanations for the same events, whether long like the Reformation or short like the outbreak of World War One, differ radically from one another. Indeed, the pattern of their succession all too often cycles and repeats itself before spinning off into an entirely new direction. Although narrative historians may be able to offer cogent explanations for their revisionism, the succession of these explanations and their lack of convergence, in stark contrast to explanations in the natural sciences, should give us pause for thought.

And there's a third thing we need to consider about history, one that should make us worry whether history is indispensable or even useful for understanding what's happening now, let alone what may happen in the future. Take the current conflict in the Middle East. All we really need to understand this conflict is what participants believe and want now, not what their parents, ancestors, and founding patriarchs believed and wanted—or even what actually happened to them—a hundred, a thousand, or three thousand years ago. But popular historians and common sense tell us it's only through history that we can figure out what motivates people now. Not just world history, but personal history, too. That's why reading people's biographies can help us understand their current and future choices. It's what William Faulkner was getting at when he wrote in *Requiem for a Nun*: "The past is never dead. It's not even past" (Faulkner, 1951, p. 80). All this is so obvious, it seems hardly worth mentioning as a justification for paying attention to history.

But this rationale for studying histories and biographies should be troubling—if for no other reason than they don't tell us what actually happened in the past, only what people think happened in the past. It's people's beliefs about history that motivate, not the actual historical events. So, even if we get the facts right, that may be irrelevant to understanding people's present or their future, for that matter. If we want to understand the present and the future, we better figure out what people now believe about history instead of what actually happened in it. But even that premise is not one we should accept without demur.

Contemporary social and behavioral sciences certainly don't vindicate the notion that people's beliefs about history—whether accurate or mistaken—are indispensable to understanding their affairs. Take economics, for example. There's almost no narrative history in most of the influential economic models of human behavior—the ones that won the Nobel Prizes for economists like George Akerlof and Thomas Schelling. Economics explains events by showing how human choice is driven by current expectations and current desires about the future. That's why Lowenstein's history

America's Bank is not going to be on many syllabi for courses on monetary theory. To understand the central bank, you need to understand the effect of reserve requirements, how buying and selling bonds affects the discount rates, and the quantity theory of money. You don't need to know anything about Carter Glass, Nelson Aldrich, and Woodrow Wilson, the men who, in Lowenstein's history, made the Fed.

In fact, the relationship between economic understanding and historical understanding is exactly backward in many cases. Do you want to understand the history of slavery's expansion in the United States, for example, or the decline of Spain as a world power, or why Britain adopted free trade in the nineteenth century? Most economists will tell you that what you need is an *a*historical economic theory. History only provides the events to be explained and the events that test the explanation.

These reasons for skepticism about narrative history's powers to confer real understanding raise an interesting question for psychology: why is it that we prefer narrative histories or stories as the most effective means of conveying information? One thing that makes this question worth asking is that, when scientists communicate their research to other scientists or teach their science to students, they never use history to do so. They hardly ever mention it. They don't think history confers understanding of what they do. But scientists and science writers all agree that most nonscientists don't see things the same way. Stephen Hawking resorted to history to explain cosmology, even though his understanding of the matter as a physicist is quite ahistorical. Bill Bryson employed history to explain everything science reveals, probably because that's the only way that he, along with most of us, can understand most things, not being scientists ourselves.

So why do science writers employ the historical strategy when they try to convey science to nonscientists? It's not enough simply to note that people like stories more than laws of nature, that they prefer narratives to formulas, that they can make more sense of movie plots than plots on graph paper. We need to know why they do.

We humans have an insatiable appetite for stories with identifiable heroes, the tension of a quest, obstacles overcome, and a happy (or at least emotionally satisfying) ending. Science writers know that if they can find features like these, nonscientists will stay interested even when they don't really understand much about the science itself. At their best, such plotted histories of a scientific achievement may convey some of the science in ways that enable nonscientists to understand more than they could glean from journal articles or textbooks, or even from conversations with the scientists themselves. Of course, it's not just the way science writers keep readers interested. Scientists, even Nobel laureates, succumb to the satisfaction, the pleasure, the release, sometimes even the catharsis of stories that explain the attractions of history to everyone else. Some write best sellers like James D. Watson's autobiographical *The Double Helix* (Watson, 1968), and everybody, including scientists, reads them for the story. It's everyone's preferred mode of understanding.

Our preference for, attraction to, and ability to remember stories are facts about human psychology that need scientific explanation. There are powerful motivations for uncovering the sources of our addiction to stories. First and most obvious, understanding the springs and sources of our attachment to narrative may make us better able to harness it in the service of other human needs and aspirations. Social psychologists share an interest with marketing executives and movie producers in discovering exactly how it is that stories satisfy in the way nothing else does. Knowing why would mean more effective advertising messages, political campaigns, screenplays, and so on. It's not just science writers seeking a place on the nonfiction best-seller lists who should care about knowing why narrative has such a hold on us. With good answers to this question, we might be able to improve science education, at all levels, even the communication of information among scientists themselves. At a minimum, we'd be able to more reliably identify barriers to broader scientific understanding.

But, as the rest of this book will argue, there is another, even more compelling reason to answer the questions raised by our love of narrative and consequent attachment to history. The real trouble with that love and attachment is that the explanations of narrative history get almost everything wrong, and the consequences are more often than not harmful. Narrative history is almost always wrong in a way that science has managed to escape. It's wrong even when a narrative gets the facts of what happened exactly right, without adding things that didn't happen, or leaving out crucial things that did.

To actually convey understanding, a historical explanation has to get its dates right. Then it has to get the causal connections between the events in the chronology right. But getting this latter right, as we'll see in later chapters, almost never happens in narrative history or biography. The short answer to the question "Why not?" is that no one has figured out a reliable theory that identifies both the real causes and identifies what it is about them that makes them the real causes. Without such a theory, even if you got the "dots" right, you wouldn't know what the connections between them were, and for the same reason, you wouldn't even know why they were the right "dots"—the significant events in the historical chain that does the explaining.

There is a telltale sign history doesn't connect the dots the right way every time we commemorate a major historical event. A hundred years after the onset of World War One, we think we have a good handle on the events that led up to it. But the spate of books published on the centenary of that cataclysmic war still disagree radically about the "right" narrative of the war—what caused it, the arms race, fears of encirclement, nationalism, colonialism, commercial competition, some of the above, all of the above, none of the above? When the war broke out in August of 1914, a former chancellor of Germany asked the then ruling chancellor why it had happened. His reply, "Ah, if only one knew," still applies. The persistence of radical disagreement about even the most well documented and most interesting events is a symptom of the problem historical explanation faces. As you'll see, the solution to the problem requires us to give up narrative history as a source of understanding.

It will be difficult to do so, of course. History is vastly entertaining. It gives us too much pleasure (along with a host of other emotions and feelings). But that's just another symptom of the problem narrative history faces as a source of knowledge and another reason it's a dangerous substitute for knowledge.

The stories popular historians tell scratch the itch of curiosity, they satisfy our felt psychological need to know why something happened, and they do so even when they are completely wrong, so long as we don't know that they are. There's no difference between the satisfaction of curiosity Robert Southey's 1813 biography of Lord Horatio Nelson provides us and the feeling we get from reading Patrick O'Brian's suite of great novels about Nelson's contemporary, Jack Aubrey, R.N., Admiral of the Blue. Curiosity assuaged is no substitute for real understanding, and no mark of it either. But when a good story puts an end to further inquiry by sapping our desire to learn more, it can be an obstacle to real explanation. Conspiracy theories of history exercise their grip on credulous people largely because they "make sense" of events in ways that exploit the desires and the ignorance of the gullible and the suspicious. The difference between these stories and the ones reasonable people believe is not in the way they satisfy curiosity. The psychological closure that both produce is an obstacle to further inquiry.

Combine the psychological effectiveness of good stories in putting an end to curiosity with the way in which they can motivate action, and you have a powerful further reason for trying to understand how stories work and whether they can convey real knowledge. Our need to figure out how stories work will be especially compelling, if as I have suggested, by and large, the impact of (spoken, written, filmed) history on (the actual course of) history has been destructive and, at times, horrific.

The suggestion that every explanation in narrative history is wrong seems far too radical for anyone to take seriously. Indeed, so radical that serious readers, and especially lovers of history, will be strongly tempted to treat the whole idea as unthinkable. In a sense, they will be right. We really can't shake our attachment to historical narrative. But learning from cognitive psychology, evolutionary anthropology, and neuroscience exactly why we love stories so much is enough to establish why we need to give them up as sources of knowledge.

For the more than just skeptical defenders of narrative history, chapter 2 will present narrative history's strongest claims on our attention as explanatory knowledge. We'll examine its own explanatory aims, what popular historians think constitutes real understanding, and, for the sake of argument, we'll accept their self-imposed standards of explanatory success as appropriate. Then, in the chapters to follow, we'll explore why historians and nonhistorians alike have always accepted these standards—but why they can never be achieved.

2 How Many Times Can the German Army Play the Same Trick?

By the summer of 1870, Otto von Bismarck had managed to play the chess game of European politics so brilliantly that he had all but created a single nation out of hundreds of German-speaking principalities, duchies, free cities, and kingdoms that stretched across the European plane from France to Russia's western province of Poland. To effect Germany's final unification into the Deutsches Reich, he contrived to involve all these separate states in a conflict with France by tricking Napoléon III into declaring war on Prussia. The Franco-Prussian War allowed Bismarck to call upon an alliance of all the German states to defend their loose confederation. Under Prussian Field Marshal Helmuth von Moltke, the German army surrounded the French and defeated them at the Battle of Sedan on the 1st of September 1870, when it captured Napoléon III. The war dragged on for another five months of desultory fighting until Paris was besieged, after which Bismarck arranged for the king of Prussia to be crowned as "Kaiser"—emperor of a single united German nation—at the Palace of Versailles outside Paris.

But the Battle of Sedan is the event that military historians remember most. After its spectacular defeat in that town, the French army established the Grande Quartier-Général—a general staff like the German one. In the forty-four years that followed, French military historians subjected the Franco-Prussian campaign and its decisive moment at Sedan to careful study, for the now unified German Empire had become an even bigger threat to France than the dozens of German principalities had been before the war.

Sedan sits in the department of the Ardennes, near the Belgian border, to the west of the small nation of Luxembourg, in the forest and mountain

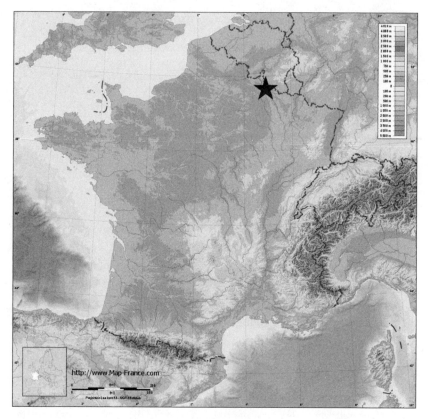

Figure 2.1
Location of Sedan on the map of France. Courtesy of Creative Commons Wikipedia.
http://www.map-france.com/town-map/08/08409/mini-map-Sedan.jpg

range that gives the department its name. It's the star in the map shown in figure 2.1.

By the late summer of 1914, "the lamps were going out all over Europe," in the words of British Foreign Minister Edward Grey. But not in the Grande Quartier-Général of the French army nor, for that matter, in the Großer Generalstab of the German army. Each was war-gaming the opening moves of the war they were both hoping for. The Germans had been refining what was known as the "Schlieffen Plan," named after Alfred von Schlieffen, the chief of the General Staff who drafted it as early as 1905. The plan was famously summed up in the admonition "When we march into France, let the last man on the right brush the English Channel with his sleeve." This time, France was to be defeated, not by a frontal attack from the west toward Sedan as in 1870, but by a vast encirclement. The German army would sweep down across Belgium along the north coast of France and then take the French army from the rear as it rushed east toward the Rhine.

But, perhaps fearing that the French knew about this decade-old plan, in the summer of 1914, the German Generalstab decided to modify it. Instead of heading west to the Channel, the German army would go through the Ardennes, almost exactly where it had met and defeated the French army at Sedan in 1870.

Despite careful historical study of the Battle of Sedan, including the terrain and the German strategy in 1870, this battle went no better for the French in 1914 than the first one. Forced to retreat almost all the way back to Paris, the French army was saved 50 kilometers from the city by reinforcements, some crucially delivered in taxi cabs (whose drivers ran their meters).

On the French battle map of that offensive (figure 2.2), you can find Sedan in the upper middle, around which the advancing German armies are sweeping counterclockwise.

In the years following World War One, a new generation of French military historians working for the Grande Quartier-Général, studied these operations assiduously, though all the analysis was kept secret.

Despite two previous rounds with Germany in the seventy years before, the French were surprised in May 1940 when the Germans did the same thing for the third time in as many wars. Hitler sent the Wehrmacht's Army Group A straight through the Ardennes right into Sedan, which was captured without resistance. The Army Group then poured south, defeating the entire French army in less than six weeks. Notice the large dark gray arrow in the map shown in figure 2.3, pointing right at Sedan.

Four years and six months later, the armies of Britain and the United States stood pretty much where the French armies had been in May of 1940 (the thick black line in the figure 2.3 map). At which point, Hitler launched his largest and last full-scale offensive of World War Two. He did so in the very same place where the Germans had attacked and won in 1870, 1914, and 1940: the Ardennes. The Allies called it the "Battle of the Bulge" for

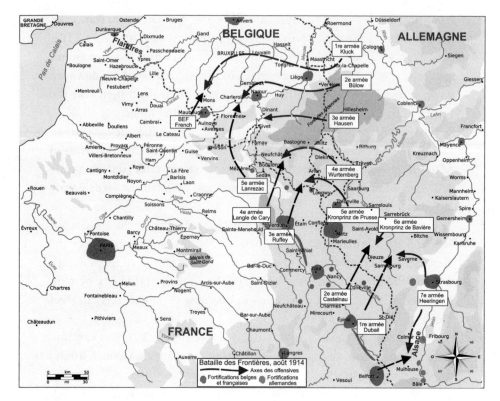

Figure 2.2

French map of Battle of the Frontiers (Bataille des Frontières), August 1914. By Lvcvlvs—Own work, CC BY-SA 3.0, https://commons.wikimedia.org/w/index.php ?curid=29123350.

obvious reasons. Notice how the Germans pushed a huge bulge into the British and American lines between the middle of December 1944 and the beginning of January 1945 (figure 2.4).

If you're looking for Sedan (whose location isn't labeled on the map in figure 2.4), it's just to the left and below Neufchâteau on the lower left of this map.

Four times in seventy years, the Germans attacked in exactly the same place. How could their adversaries get matters so completely wrong so many times in the same way at the same place? Common sense and prudence would strongly suggest precautions against the Germans doing the same thing over and over again. Well, that's what common sense might tell us.

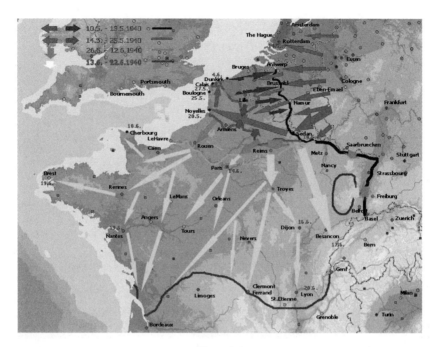

Figure 2.3
The fall of France, May–June, 1940.

But the professional military historical analysis of what went wrong each time before, in fact, seems to have suggested each time that it wouldn't, couldn't happen again.

It's important to note that the French, British, and American general staffs actually took seriously what their historians told them, and they planned accordingly. The staffs treated history as a source of knowledge useful for predicting the future. And, each time, they paid dearly for doing so.

After the fiasco of the Franco-Prussian War, France completely reorganized its military, and with careful study of the history of the war, changed its military doctrine, too. Long before 1914, the French decided that they were not going to make the mistake of defending cities under siege, as they had done in 1870. Instead, they would attack with élan right through the Ardennes on either side of Sedan. But their advancing armies were decimated by machine gun fire and artillery. Afterward, it didn't take much study of World War One for the French to realize that it is defense, not

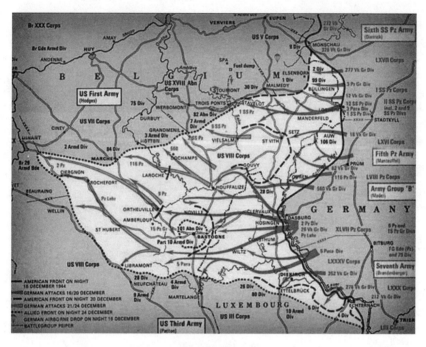

Figure 2.4
Furthest extent of German advance during the Battle of the Bulge, December 1944.

all-out offense, "à l'outrance," that wins wars. So, the third time around, they built the Maginot Line, as history bid them, to prepare for the next World War, spending billions of francs in the process. And so, too, as students of history, in 1940, the French couldn't believe that the Germans would do it again. They just knew that the Germans, unable to get their tanks, half-tracks, and other armored units through the deep forest of the Ardennes, would have no choice but to come down through Belgium, as the old pre–World War One Schlieffen Plan required them to do, and to fight it out in front of the fortified Maginot Line. They were wrong, and it cost them the war. In 1944, the Allies thought they knew exactly what the Germans were going to do because they had broken Enigma, the German secret radio codes, years before. Since there were no orders radioed to German units to prepare for an offensive through the Ardennes, the Allies concluded there couldn't be one coming. They were wrong— a mistake that almost cost them the war in the west. The Germans, suspecting that their radio traffic was being intercepted, had ordered their officers to use only telephones—and they attacked in the same spot they'd used three times before.

The moral of this story is that, if you want to prepare for the next war, don't study military history. But the military establishment of every modern Western nation devotes vast resources to have historians do exactly that. More than their colleagues in other areas of study, military historians take seriously the oft-repeated dictum of the American philosopher George Santayana. In *Reason in Common Sense*, he famously wrote, "Those who cannot remember the past are condemned to repeat it" (Santayana, 1905, p. 284). Others have seconded this sentiment. The world's most widely read historian, Winston Churchill, is widely reported to have claimed that "the farther backward you can look, the farther forward you are likely to see."

Military histories may be the most convincing evidence of the failure of history to get the future right. For centuries now, governments have paid some of their best and best-informed military officers to write the military histories of their wars. The result has been a cornucopia of studies of the minutest details in the exercise of the military art. The U.S. Army's history of World War One comprises nineteen large volumes and its history of World War Two, thirty-eight.

No one can pin down the origin of the saying "generals are always preparing for the last war," so widely known and so often repeated. But, surely, the military historians have long been acquainted with the observation it makes. You would think that they'd pay some heed to it, especially since they are expected to provide insight from history to guide the strategy, tactics, and logistics of future wars. You'd be wrong. And the problem is not that military historians are stupid or uninformed. The problem is that history is not up to the task of providing useful guidance about the future.

The story of the military historians' repeated failures to extract the right lesson from their study of the past, to see their way forward by looking back, is the clearest illustration of a more general fact about historical scholarship. There are obvious reasons why the military planners should not have looked to military history to predict the future. These reasons also apply to most other historical research. Here are a few of them.

First, the exact factors a historical explanation cites for one event almost never recur again in exactly the same way. So one explanation doesn't have much bearing on the future. Even if the same factors are in operation in the future, their weights and proportions are rarely the same, and new factors are almost always involved. This is far from news to most historians, and it is among the reasons many have dissented from Santayana's dictum and looked with skepticism on Churchill's pronouncement.

But the users of history have increasingly underestimated the role of entirely new factors in shifting the trajectory of events away from its past directions. The effect of entirely new and unarguably unpredictable factors on history has grown rapidly in the last two centuries. Ironically, it's safe to predict the role of these factors will soon completely undermine confidence in any predictive role for history. This prediction is not itself based on history, though the history of the four separate German thrusts through the Ardennes illustrates it plainly enough. Technological change from 1870 through 1944 was clearly unpredictable and had a decisive impact when added to other factors that remained the same over the period. The machine gun, the Maginot Line, the tank, Ultra decryption all made a difference in what happened in each of the thrusts—and all made any extrapolation from previous history pretty pointless. It's interesting to note that the military planners were later criticized by historians for not having learned this obvious fact from history. If technological change is driven by scientific discovery and invention—the quintessentially unpredictable results of human creativity—then it seems obvious that the events technology affects will be just as unpredictable. And, as the role played by technology in human affairs grows ever greater, the study of history will have fewer and fewer lessons for the future.

A second, equally important complicating factor in efforts to apply history to the future is the knowledge the historical "players" have of that history itself. One reason the German attack through the Ardennes was such a surprise, especially the third and fourth times they tried it, was that everyone knew they'd done it before, and no one suspected they'd do it again, and again, and again. But the Germans' decision to do it again was based, at least in part, on the conviction that their adversaries, knowing the Germans had done it before, again and again, would simply not expect them to do it yet again. History—the study of what happened and why—has an effect on history, on what actually happens, through people's awareness of it. That kind of "reflexiveness" is another source of the difficulty in extrapolating the human future from the past.

If, in general, the course of human history is heavily influenced by the emergence of new factors that no one could predict, like scientific and technological change, and if it is also "reflexive," effected by historical agents knowing what happened in the past—then it's pretty safe to conclude that history has no hope of predicting any detail that might be used to cope with the future. Some people view this conclusion as reflecting a serious objection or at least a limitation on history's claims to provide knowledge, or at any rate useful knowledge. It certainly undermines grandiloquent claims like Santayana's and Churchill's. And it provides the basis for the animus of at least some empirically oriented social scientists against history.

But, as the basis for a critique of history as an explanatory discipline, the demand for predictive power is widely viewed as being both unreasonable and a misunderstanding of what history is all about. This goes for both academic historical scholarship and the true-story narrative history devoured by general readers. Most historians and consumers of history will insist that history is not in the event-prediction business and can't be judged by any such a standard—indeed, that the goal, function, aim, and objective of history and historical scholarship is *not* to teach us anything specific about the future. Some sciences are in the prediction business and do increasingly well at it. But history isn't science and doesn't pretend to be. Don't mistake history for accurate crystal ball gazing, they insist.

Although, unlike weather forecasting, history doesn't—and indeed is not supposed to—predict particular future events, it does nevertheless prepare us for our future, or so some argue. It's more like climate science, measuring out epochs and periods, explaining events by "contextualizing" them as instances, examples of much larger units, periods, what a famous French historian called the "longue durée." As such, history is the only thing that can help us cope with the future. How else, so many historians argue, can we understand our present except as the result of our past? And although the future is open, it's not radically open, and only history can identify, if only tentatively and imprecisely, its shape. Sometimes history helps us avoid traps, mistakes, outcomes that we had to deal with in the past and wish to avoid. Sometimes it suggests ways to improve the chances of attaining outcomes we seek, or mitigate the worst consequences of ones that are unavoidable. Sometimes it can reveal trends, help us put the present into perspective, measure the dimensions and the causes and effects of events in our own times by comparisons. History can do all of this, even if it can't actually tell us what's going to happen next. Surely, that would make a knowledge of history worth having.

Well, such knowledge certainly would be worth having if there were a way of telling its right explanations from its wrong ones, and those of its right explanations which were broadly relevant to the present and the future from those which were not. Having a good idea about the direction or drift of things would be of inestimable value to our understanding if only we could be confident that we indeed had such an idea. But it's evident there's no agreement among historians, either on how in general to tell right from wrong explanations, or even which particular explanation is right or wrong, still less which ones can help us prepare for the future, whatever that means.

Each of the four military catastrophes in the Ardennes that military historians didn't predict was the subject of a best-selling work of history by a respected historian: *The Guns of August* (Tuchman, 1962), *The Fall of Paris* (Horne, 1965), *To Lose a Battle* (Horne, 1969), and *The Bitter Woods* (Eisenhower, 1969). None of these best-selling explanations, not Barbara Tuchman's, not Alastair Horne's, not John Eisenhower's stood for very long without inviting a revisionist historian's response. Why is that? Because it remains an open question whether any of these best sellers got it right, correctly explained why the Germans went through the Ardennes in 1870, or 1914, or 1940, or 1944.

There is overwhelming evidence that the most basic questions about the most significant events in history must remain permanently open. Long after the events of World War Two, indeed long after all the archives were opened, diaries read, and memoires published, historians still disagree. Between 2014 and 2016, at least eight books were published debating the events of the 1944 Battle of the Bulge alone.

In 2014, at least a half a dozen door-stopper books on World War One were published, each with a different narrative explanation of why the Great War had broken out a hundred years before. Now, you might say that the appearance of yet another batch of books on such a well-trod subject is testimony not to previous histories' having got matters wrong, but, rather, to our eagerness to consume narratives. The cause for their publication is sheer commercial opportunism, not new revelations about why World War One really broke out. Indeed. But, surely, the tide of history books on the same subjects, year after year, decade after decade, and now century after century, all disagreeing about their causes, is also testimony to the fact that we still lack the historical knowledge we seek? We still don't know why World War

One started. Worse, we will never know if any of these new books have provided the correct answer to this question.

If it's not just commercial opportunism, there must be a reason for the spate of books still disagreeing about what brought about World War One. The disagreement is still lively because, no matter what historians say, we and they seek guidance from history for the future. One historian's explanation is a cautionary tale that he or she hopes will shape policies to avoid a similar cataclysm. Another's shows that we have nothing to worry about, or at least we don't have the same thing to worry about because circumstances have changed so much. Still a third explains the outbreak as a tissue of one-off happenstances, freak accidents that have no implications for avoiding future outbreaks of war.

Unless there's a way to tell which account is right, the disagreements will never end. Now, in the sciences, we know perfectly well how to decide between competing explanatory theories, models, or simulations. The criterion is clear. It's predictive success. In choosing between alternative theories, scientists sometimes have to compare predictive track records or to examine the trends of increasing or decreasing predictive accuracy as their measuring instruments improve. But, as we've already seen, such an approach is not only not available in history; it's not even appropriate. Unlike the sciences, history is not in the business of predicting the weather or earthquakes, much less political elections, economic depressions, or violent revolutions.

But unless historians accept some kind of demand that their explanations have even a modest predictive payoff, they won't be able to claim a role for history in helping us avoid pitfalls, tracing out the broad outlines of human affairs, or enabling us to cope with the future by narrowing down its possibilities to some manageable, intelligible proportions. Without some track record of predictive success, even this role is too much to demand of history as a body of explanations we can rely on for much of anything at all.

The question remains open, what kind of real understanding do historical explanations provide if they're not to be judged by their value in shaping our preparation for the future or our coping with its challenges?

Historians, the defenders of narrative history will insist, have far more important work to do than offer predictions or even guidance about the future. The value of historical explanations can't be judged on that standard because the kind of understanding they provide is quite different.

And, once we recognize the real aims of history, they go on to argue, we'll see how it matters for the future (and the fleeting present for that matter) without making unreasonable demands on its powers to foresee anything in particular.

Historians seek the *meaning* of the past for the present and future because that is the best way, or the only way, we can understand our circumstances. And all the best of the explanations they provide help shape and ground that meaning for us.

Historians certainly do sift through the events of the past to identify the forces and factors they think had the greatest importance in shaping the present: since the present is a moving target, history must get rewritten each generation since the past factors that were significant for making the world of our fathers and mothers may not be the same as the ones making our world or our children's world.

History enables us to look back from the perspective of the present to find the real *meanings* of past events. This is a role for history that has nothing to do with predicting the future or coping with it in any way. History reevaluates, reassesses, revises the past in the light of the present, finding features sometimes unnoticed in the past but important to us today, to give the past its "real" or at least its "contemporary" significance.

The American Civil War, variously called the "War of the Rebellion," the War between the States," and the "War of Northern Aggression," is a wonderful example of the role and importance of history, even as it is revised, from generation to generation, in putting whole epochs, eras, periods, in a new light, one that reveals features important to us. It's not just the names for "the late unpleasantness" that have changed. Over the hundred and fifty years since the war's end, each generation of Americans has sought a new import in the events of 1861–1865. The meanings adopted in different times and different parts of the country have differed quite sharply, and the differences have been quite powerful in their effect on culture, society, and government. Historians and others have argued over the matter for 150 years or more. The war came to mean several different things over time in large swathes of the United States. For a hundred years or so, the Southerners' interpretation prevailed: the war was not really about slavery at all, but about states' rights. Whence the name "War between the States." Slavery as a fact about the South, "the peculiar institution," became invisible as a *casus belli*, and the Northern victory was fatally weakened as a motive to oppose Jim

Crow segregation. For a century, Southern and Northern historians perpetuated the interpretation of the Civil War as the culmination of an argument about "nullification" and "interposition," not an argument about the extension of chattel slavery. It thus could not take on the meaning, significance, and moral force that would make it a motivating narrative for opponents of racism and exploitation.

The rewriting of the history of the American Civil War in the last twenty-five years of the twentieth century completely changed its meaning. Looking back, a changed contemporary America found a new meaning of the Civil War in the abolition of slavery and the sacrifices of African American soldiers. This new meaning of the war didn't result from the discovery of new demographic data, new sources in diaries, letters, and memoires, new quantitative studies of slavery as a system, but, rather, from the new importance assigned to these, which motivated the new historical explanations for the Civil War and for newly important incidents in the war. It made possible and brought about a whole new domain of social history to complement, displace, deepen, and change the military and political history of the war. It made visible the role of many in the war who had been invisible, black Union soldiers, slave and free women, Southern opponents of slavery, Northern apologists for it. The change in the war's meaning for us challenged and often overturned the weightings of importance assigned to the causes of the war on which almost everyone had long agreed.

Notice, however, that the sea change in the meaning of Civil War history over the lifetimes of most Americans was more a source than a result of the revisionist history that has flourished since the centenary of the Civil War in 1965. James McPherson's *Battle Cry of Freedom* (McPherson, 1988) reigned over this tide as a unifying document. Social, economic, cultural, and gender, along with technological, military, and political, historians all contributed to filling out the details of what happened in the United States from before to well after the period 1861–1865 in ways that would vindicate the meaning of the war as part of an emancipation for all Americans. (British readers may find the same shifting sands in the history of the meaning of the English Civil War, and French readers in reinterpretations of the German Occupation of the 1940s.)

A large part of the project of giving meaning to past events, periods, and epochs is to be found in its moral motivation. Although nothing makes this plainer than the popularity of the labels the "Good War" for World War Two and the "Greatest Generation" for those Americans who endured the Depression of the 1930s and went on to fight the Good War, both these labels are contested. Was it really a good war, or at least a morally better war than the others the United States has fought? Was the generation who experienced it really morally more worthy than the generations who came before or after?

Much narrative history is written and read in order to vindicate or contest such moral judgments. This, many historians and many readers of history will maintain, is why history is so important, why its explanations matter. What doesn't matter is whether the explanations of history have any kind of payoff for helping us cope with the future. Their importance lies in how they enable us to evaluate the past, plumb its meaning, significance, import, its weight for meting out reward, punishment, praise, responsibility, blame, or exculpation, even if these determinations are entirely personal to the writers or readers who make or subscribe to them.

Notice, this role for history still makes getting its explanations right of the highest importance. For without confidence that among competing historical explanations some are right and others wrong and that we can tell which is which, history amounts to little more than propaganda or ideology deployed in conflicts over values, interests, and power.

Recall the controversial claim of chapter 1, that history as an account of the past in human affairs has been more a force for harm than good (to which we'll return in the last of the main chapters). Perhaps the most powerful response to that charge is to argue that history is or would be a force for good or at least for better when and only when its narratives get their stories right. The right explanations shared widely enough can make history largely, perhaps even entirely, a force for good.

This positive view of history's impact when it gets matters right may be overoptimistic. But surely getting the narrative right is a necessary condition for historical understanding. McPherson certainly holds this view. He ends his magisterial history of the American Civil War by stressing the importance of getting the narrative right for the broadest historical understanding: "Northern victory and southern defeat in the war cannot be understood apart from the contingency that hung over every campaign, every battle, every election, every decision during the war. This phenomenon of contingency can best be presented in a narrative format—a format this book has tried to provide" (McPherson, 1988, p. 77).

Historians have a responsibility to get their explanatory facts right, and not just because history claims to provide knowledge of the past, but owing to the impact of its explanations on the rest of humanity's concerns. The demand that historians get these facts right, that their narratives uncover and identify the real causes of what they purport to explain, immediately poses the problem of how to tell when they have done so. We've excluded predictive success, even of the most generic and open-ended kind, as the criterion of whether and when historical explanations are right or even moving closer to the truth. What will provide the litmus test we need?

Historians will have little patience with this challenge. After all, we know in the ordinary course of life and from our daily interaction when the explanations we offer to one another are right or wrong, credible or incredible, worth relying on or mere rationalizations or excuses. We know perfectly well what kind of evidence will strengthen our confidence in their rightness and what sorts of facts will weaken them. The same goes for historians. Of course, they have a much harder job than ours is in ordinary life. And the more distant the past, the harder it is for them to find the sort of evidence that will determine which narrative is the right one for what actually happened. But that's no reason for general skepticism. Sometimes the evidence that archival scholars presents to rule out or rule in a historical explanation is even better than the evidence that law courts require. Historical narratives can get their explanations right, many have argued, and we can tell when they do.

It is this widely held, indeed almost universal, view that this book will challenge. It will show that historical narratives are wrong—all of them. Even when most people can't tell they're wrong, neuroscience reveals why history's narratives are always wrong. If we seek real knowledge of the past, we are going to have to rely on something quite different from what most people rely on to actually find such knowledge. This will be the case whether we look to history for some practical payoff in coping with the future or seek to understand what happened in the past for its own sake.

We'll start, in chapter 3, by considering the best narrative explanation of an important event whose occurrence has long perplexed many.

3 Why Ever Did Hitler Declare War on the United States? That's Easy to Explain, Too Easy

World War Two history buffs often ask themselves, Why did Hitler declare war on the United States (figure 3.1)? In December 1941, he had his hands full with the Soviet Union, and he hadn't even defeated Britain. What's more, after the attack on Pearl Harbor, the United States was so focused on Japan it probably wouldn't have joined in the war in Europe anyway. The Germans might have won.

Here's one of the best historical explanations for the German declaration of war, which brought an unwilling America into the war against it:

[Hitler] felt a rush of blood after Pearl Harbor. Neither he nor anyone else in the Nazi leadership had *anticipated* an attack of such boldness. The very audacity of the Japanese strike *appealed* to him. It was his sort of move. And, grossly overestimating Japan's war potential, he *thought* its effect was far greater than it turned out to be. In those days reeling from setbacks on the eastern front (the first devastating Soviet counteroffensive of the war had just begun), he could not have *wished* for better news than a Japanese assault on the American fleet at anchor. Japan and America at war was exactly what he *wanted*. The *decisions* that followed were taken in a mood of exhilaration. But they were not driven by spontaneous irrational emotion. Letting his U-boats loose on American shipping came first. He had no doubt been itching to do this all autumn. Now he need hold back no longer. This in itself, he *imagined*, would turn the battle of the Atlantic Germany's way (and indeed a small number of U-boats at work off the northern American coast were able to wreak havoc on Allied shipping in early 1942). It preceded the bigger *decision*, to declare war on the United States. Prestige and propaganda considerations dictated that this should come from Germany, and that he should not passively await a declaration by America. But Hitler's own *decision*—and, as we have seen, it *was* his, taken without consultation apart from with subservient [Foreign Minister] Ribbentrop, and presumably [Generals] Keitel and Jodl—had been preceded by moves, *rational from his point of view,* and dating back several weeks, to prevent Japan, once in the war, from leaving it at a time that did not

Figure 3.1
Hitler declares war on the United States, December 11, 1941. https://
en.wikipedia.org/wiki/German_declaration_of_war_against
_the_United_States_(1941)#/media/File:Bundesarchiv_Bild_183
-B06275A,_Berlin,_Reichstagssitzung,_Rede_Adolf_Hitler.jpg

suit Germany. Only when what effectively amounted to a new tripartite pact [Germany-Japan-Italy] had been concluded did Hitler declare war....

Germany's role, in supporting Japan by entering a war against America which, to Hitler, was inevitable anyway, was to prevent the Americans defeating the Japanese or forcing them to agree terms, before turning on Germany. The United States, through German intervention, would be forced into a war on two oceans....

From his perspective, therefore, the declaration of war in December 1941 was no great gamble, let alone a puzzling *decision*. He felt he had no option. The *decision* seemed to him to open up the path to victory which was by the beginning of autumn 1941 to recede. For him, therefore, it was the only *decision* he could make. (Kershaw, 2007, pp. 425–426; emphasis added)

Now you see why the Germans lost the war. This is the kind of knowledge history is supposed to provide, not reliable predictions but retrospective understanding.

What should be puzzling about this historical explanation is why you understood it, and why you would have understood it (couched in suitable vocabulary) even when you were a small child. The explanations of narrative history work without adding anything by way of theoretical machinery, conceptual apparatus, special stipulations, or explanatory strategies, devices, or models. In this, they are unlike explanations almost anywhere else. They require no special gifts to fully grasp them. The fact that everyone loves historical explanations, that loving them doesn't require any special preparation, unlike explanations in math or physics or economics, turns (at least in part) on the fact that everyone immediately understands them, and from earliest childhood.

Think about the subjects you studied in college. In many cases, you had to learn some basic theory right at the beginning: when you studied economics, for example, you had to learn its unfamiliar laws and concepts—concepts such as "real prices" (which aren't given in money terms at all) and "ordinal utility" (which has little if anything to do with actual needs or wants). Or when you studied French, you had to learn which verbs take "avoir" and which take "être" in the "passé composé." And when you studied other subjects—most notably biology—you may have had to *un*learn some mistaken theory you were taught in high school. Or mathematics, where you never had to unlearn the theory you'd been taught before, but where you may have found the theory you were now taught so hard you forgot it almost as soon as you learned it. Even appreciating literature required you to learn at least some literary "theory." The only subject that you could dive straight into without learning any new theory or, for that matter, any theory at all, was narrative history. From the very beginnings of our education in the subject, whenever it began—in kindergarten, at our grandmother's knee—none of us ever had to first learn a bit of theory.

That means one of three possibilities: either (1) understanding narrative history doesn't require any theory; (2) we learned that theory as infants without realizing it, even before we began listening to stories about the past; or (3) we knew the theory at birth—innately.

There is a pretty compelling argument for ruling out possibility 1, the "no-theory-necessary" possibility of how we understand narrative history. A list of events, even one in chronological order, doesn't explain anything. To explain something, the events need to be connected as links in a causal chain or network. But, as David Hume pointed out in the eighteenth century, there is no observable causal linkage between events, whether in nature or in human affairs (Hume 2007 [1748]). Finding the causal linkage that holds events together is therefore a matter of guesswork. We impose, improve, and correct those guesses by dint of experience and observation. At their best, these guesses about the causes underlying the great theories (hypotheses really) of science, like those of Isaac Newton on gravity, Charles Darwin on natural selection, and Albert Einstein on the curvature of space, are guesses about "causal linkage." Some theory or other is indispensable in understanding anything because it turns a mere sequence of events into a causal chain that can explain the links.

This is as true in narrative history as it is in natural history, where there is really only one basic theory that guides the search for the right chain of events: Darwin's theory of random or blind variation and natural selection. So there has to be a theory, or perhaps more than one, that does the same thing for human history—that makes historical narratives understandable.

This theory has got to be one we employ automatically and unconsciously, although it's not the only such theory we employ without conscious reflection. Many of the complex cognitive activities we engage in require unconscious employment of very complicated theories. Indeed, in some of these activities, we can employ the theory effectively *only* if we do so unconsciously. For example, deciphering sounds into meaningful speech and encoding thoughts into meaningful sounds in our native language require us to employ the grammatical and semantic theory that constitutes our spoken language without consciously thinking about it. Indeed, if we had to employ this theory consciously, as we do when learning a second language, we wouldn't be able to hold a normal conversation. Much the same is true of the theory of mind. In almost all cases, we employ it unconsciously. But the fact that we use it without realizing it, unconsciously, makes it invisible to us.

Since we didn't learn any such theory in school, that leaves the remaining two possibilities noted above, which psychologists have been arguing about for fifty years or so: did we learn the theory in infancy? (possibility 2) or did we already know the theory at birth, innately? (possibility 3). It has to be one or the other.

Let's put off to chapter 4 the question of whether narrative history theory is hardwired in us or we learn it soon after birth. For our immediate purposes, it will suffice to agree that the theory has been with us since before we can remember, that it's very hard to notice, if at all, and that we employ it automatically and unconsciously. And if the theory isn't innate, then it's certainly a theory that human evolution has strongly primed us to acquire and one that human culture has universally taught.

What is this theory that creates narrative histories out of what we humans have done and what we've said across space and time? For that's what the theory does: it takes as inputs the observable data of human behavior—how we've acted and what we've said—and then it provides hypotheses about what motivated and guided our behavior in order to explain it. If we're lucky, it can sometimes even generate good guesses about what other people will do next.

Psychologists and philosophers of psychology have various names for this theory. Among psychologists, it's called "theory of mind." Some clinical psychologists, especially in Great Britain, have used the term "mentalizing" to describe employing this theory to predict and explain the behavior of others. Philosophers of psychology, for their part, have adopted the term "folk psychology" to describe the hypotheses we employ, consciously or not, to explain and predict the behavior of other people. We'll use the label "theory of mind."[1]

---

1. There is an "in-house" dispute between some cognitive scientists and philosophers of psychology about how we use the theory of mind. Some argue that we apply it only to others, and some argue that we use it to see what we would do if we were others and then we infer that others would do likewise. In this book we'll ignore this dispute since it makes no difference either to stories or to narrative history.

To a first approximation, the theory of mind everyone carries around and uses to explain and predict the behavior of others is so simple and mind-numbingly obvious, it seems almost preposterous to call it a "theory." Stating the theory is just a matter of offering a lot of platitudes, claims so obvious and trivial that we would no more dispute them than bother to mention them in the first place. These platitudes are about desires, beliefs, actions, how they are related to one another and what their causes are. Here are some of them:

> People have beliefs, thoughts, expectations.

> People have desires, wants, preferences, hopes, antipathies, fears, dislikes.

> People have sensations—they experience colors, smells, tastes, textures, pains, and pleasures—which shape their beliefs.

> People have emotions—shame, guilt, anger, disgust, happiness, sadness—which shape their desires.

> People act on their beliefs about the world in order to attain their desires.

One way to see that these platitudes are part of a theory we carry around with us and use, the way we use other theories, is to draw a diagram of roughly how the theory of mind works. The diagram in figure 3.2 is from the work of a well-known cognitive scientist, Shaun Nichols (Nichols et al., 1996). Of course, Nichols's version of the theory is more explicit than the theory we carry around with us, and its parts are labeled in more technical terms; the figure 3.2 diagram reflects these "improvements," which are needed to explain the theory and how we put it to work.

The fundamental idea of the theory of mind is that in our own minds and in everyone else's there is something like a "belief box" into which come one or more particular beliefs, stored elsewhere or newly acquired from "perceptual processes" (sight, hearing, smell, taste). Some of the beliefs in the box may also come from inferences we make by combining other beliefs we store in our memories. Many of these inferences, bringing forward beliefs we store in memory, are unconscious, as some or all of the beliefs in the belief box may be at any given time. There's also a "desire box," filled by one or more desires—long standing or immediately acquired, but delivered from some bodily feeling—whence the "body-monitoring system" lozenge-shaped box in the diagram. The desires in the desire box and the beliefs in the belief box are not just random pairings; the

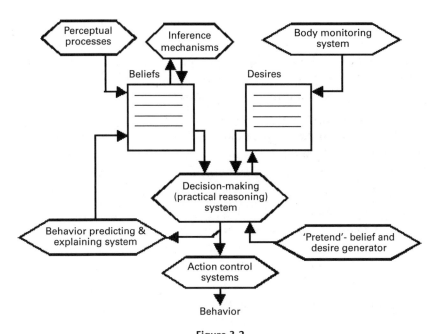

Figure 3.2
The "Boxology" of the theory of mind. From Nichols et al., 1996, fig. 1.

desire and the belief in each pairing have found their way to their respective boxes because they are "related" to each other in a way that produces the choice, the decision, the action they explain. How are they related? To a first approximation, the relationship between these paired beliefs and desires is roughly that of ends and means—the desires in the desire box provide information about the ends or aim of the action to be explained or predicted, and the beliefs in the belief box provide information about the means available to attain the aim. Of course, beliefs can be mistaken and desires unattainable, but these are complications we can ignore in sketching how the theory of mind works in general.

It was the Greek philosopher Aristotle who first noted the nature of the relationship between beliefs and desires, though everyone who has reflected on the matter even a little has already grasped it. He called the relationship the "practical syllogism," suggesting that a belief and a desire work together like two pieces of syllogistic reasoning. Instead of a statement, they imply an *imperative* that describes an action to be done as a necessary consequence of the two "premises," or, in our theory-of-mind schema, the necessary consequence of the belief in the belief box paired with the desire in the desire box. Whence the label "practical reasoning" in the diagram.

The decision-making (practical reasoning) system uses something like a principle of rational choice: given the desires in the desire box, it chooses the best means to attain them on the basis of beliefs in the belief box. The decision about what to do that results from combining the contents of the belief and the desire boxes moves us to take some action, anything from making noises and gestures all the way to moving our entire bodies. It can also make us cease doing something or refrain from doing something. That's what the "action control system" is all about. It guides our behavior in accordance with the results of the practical reasoning.

Here's a simple example of how we employ the theory of mind to explain behavior. Consider the man next to you at the bus stop. Why is he carrying an umbrella? Hypothesis: he believes it's going to rain, and he wants to stay dry. Putting those two items into the man's belief and desire boxes at the moment, you use the theory to crank out an explanation in the component of the theory labeled "behavior-predicting and -explaining system," which sends the output back to the belief box, adding to your stock of beliefs about the man with the umbrella. In particular, you now believe that the combination of the belief and the desire caused his taking an umbrella when he left the house this morning. Notice this isn't the only hypothesis that explains his behavior. Here's another: he wants to look like a British banker, and he believes that British bankers carry umbrellas even when it's not going to rain. A third, equally far-fetched hypothesis is that the man wants to hide a limp and believes that an umbrella will work like a cane. You can make up many more such belief-desire pairings: the man believes the umbrella is broken and wants to have it repaired, or he believes he borrowed it and wants now to return it, or he took the wrong umbrella home from a restaurant and wants to exchange it for the right one. The only limit to these alternatives is your imagination. All of these hypotheses could have been generated by the "pretend" belief and desire generator in the diagram that feeds into the decision-making system, and would also explain the man's behavior. Each hypothesis follows from a belief-desire pairing that might be in the man's belief and desire boxes. Each pairing differs from the others about why the man is carrying an umbrella, but they all employ the same theory of mind that actions are caused by some kind of means-ends combination of beliefs and desires. Because of the way the belief in the belief box and the desire in the desire box are related, they not only bring about behavior; they also make sense of it. The belief plus the desire somehow make the man's behavior "intelligible" in light of what the man wants and what he believes. They "make sense of" and "make intelligible" the man's behavior—but to whom? Well, obviously in this case, to you.

We use the theory by first trying to figure out what people believe and want, and then asking ourselves what we would do if we had those beliefs and wants. If we'd do roughly what they did do, then their actions make sense to us, we understand them, they are sufficiently well explained that our curiosity about why they so acted is satisfied.

When our best guesses as to what people believe and what they want fail to make their choices sensible, intelligible, "meaningful" to us, we have to go "back to the drawing board" and figure out why. For example, when we see a person harming him- or herself or causing pain to close kin or destroying things that person worked hard to acquire or create, we are puzzled. When this happens, we usually do one of two things: Either we hypothesize that the person's beliefs and desires are bizarre or otherwise aberrant, and we may not be able to fathom them by additional inquiry. Or we decide that the person is mentally ill, that his or her belief-desire machinery is not functioning normally. In the first case, we operate using our theory of mind by swapping out either what we had assumed to have been put into the person's desire box by the body-monitoring system or in the person's belief box by perceptual processes or inferences from other beliefs. Then we crank these new beliefs through the "practical reasoning" decision-making system to produce the behavior we have observed. When even this won't work because the beliefs and desires we think the person actually has can't produce that person's behavior, we infer that something is interfering with the processes labeled by the arrows in the theory-of-mind diagram, that there is something wrong with the mind we are theorizing about, that there is a breakdown, a mental illness.

How do we employ the theory of mind in prediction? To use it to predict the behavior of another person—friend, foe, stranger, family member, child, adult—what we need to do is fill in that person's belief and desire boxes. The simplest way to do that is to assume that the person, a woman, for example, sees, hears, smells, feels, tastes, the same things we do, stores roughly the same information from her processes and has the same immediate wants, desires, fears, likes, dislikes, and so on that we do. Or that she has the same perceptual information and motivational states that we would have if we were in her shoes, saw what she saw, wanted what she wanted, and so on.

All this is still so obvious it seems preposterous to call it a "theory." Not only that, but if we try to state the theory explicitly, there are so many obvious exceptions to each of the theory's platitudes, it's hard to formulate any one of them in a way that doesn't make it obviously false. We all sometimes fail to do what our beliefs and desires together should, according to the theory, cause us to do. We all make mistakes, we all forget, we all miscalculate, all sometimes have conflicting desires and inconsistent beliefs, which lead to ineffective and sometimes irrational actions. Any very precise version of the theory of mind will have to reflect these sources of "breakdown." Even listing all the ways the theory can break down is probably impossible—there are simply too many of them. Trying to state the theory explicitly so that it has the precision and qualifications a scientific theory requires is a fool's errand. But that doesn't mean that we don't use the theory. It means only that the full theory, the one we use, is sophisticated, complex, full of details and qualifications. Moreover, we learn to use it with great subtlety, nuance, and imagination over the course of our lives.

The words "belief" and "desire" that we use to describe the two boxes in our model for the theory of mind have many synonyms and near synonyms, and there are other words that describe particular kinds of beliefs, desires, and combinations of them—"expectations," "fears," "hopes," "worries," "plans," "intentions"; the list goes on, and each of these words can be defined in ways that reveal its component beliefs and desires. Moreover, there are different kinds of beliefs and desires: there are the ones that seem to be explicit in consciousness because we're thinking about them: for example, wanting to get to a meeting on time, and believing we'll have to speed up if we're to make it on time. But there are many beliefs we act on, without actually having them before our minds in consciousness. For example, when you look at your watch to gauge how fast you'll need to go to be on time, you don't explicitly recall beliefs about the names of the numerals on the digital readout of your watch. You don't actually recall the nature of your desire not to be late. Is it merely that you want to be considerate? Or that you don't want the other party to worry? There are vast numbers of desires and beliefs we all carry around in our heads and never bring to consciousness. Here are just two examples so obvious they're silly. We all believe that there were no Boeing 747 jumbo jets during the lifetime of Julius Caesar, for one. And we all believe that the White House and Buckingham Palace aren't made of merengue, for another. And though neither thought ever occurred to you, you believe both of them, right? We have a lot of these "nonoccurrent" beliefs. Why call them "beliefs"? Because if someone asked you, "Do you believe that Buckingham Palace is not made of meringue?" you'd say, "Yes." We carry around an enormous number of these beliefs in our heads. How we do it is an interesting question, but one we won't address. We also carry around an equally large number of desires, wants, hopes, fears, and so on. If given the choice, for example, almost all of us would desire to win a lottery that paid out, say, $4,358,267,912, rather than one that paid out $6,782. We could substitute numbers into this statement indefinitely, producing different desires we all have and never thought about.

The theory of mind makes use of both "occurrent" and "nonoccurrent" beliefs and desires (those which have occurred to us before and those which never have) to explain and predict behavior. It assumes also that people's memories for facts about their surroundings, past and present, enable them to meet most of their desires, and that they can put together a lot of what they have learned but don't explicitly remember in occurrent beliefs to guide their actions to attain their desires. The same goes for many desires that don't even occur to us—ones that don't present themselves explicitly in conscious experience.

Guessing what's inside people's belief and desire boxes is difficult. At least, it is today. It was probably a lot easier back before the Holocene, when the last ice age ended and human populations began to settle down and complicate their lives beyond hunting and gathering. If you'd been alive at that time, you could have figured out what other people desired and believed just by looking around. Their beliefs were largely about the same environmentally salient facts that your sensory organs enabled you to take in—the weather, the state of the fire in the fireplace, the sharpness of the hand ax lying next to it, the location and disposition of any nearby predators and prey, the whereabouts of your mate and offspring. And their desires were pretty easy to guess, given their behavior and given your knowledge of your own desires. Most members of a small hunter-gatherer clan were hungry and thirsty at the same time, shared an immediate desire for warmth or shelter. We humans were already carrying around a pretty sophisticated theory of mind in the Pleistocene.

Employing a theory of mind becomes increasingly difficult as the range and detail of what we have beliefs about and desires for increases, that is, as human culture and its artifacts, norms, practices, institutions expand. Nowadays, people have such a wide range of beliefs and desires regarding so many different things, that it's much harder to guess what exactly is in their belief and desire boxes at any one time. The possible range of inputs to the belief-desire model has increased so much, it's much harder to use the model to predict what someone will do and to explain what someone did with any real confidence that your explanation is right—that it identifies exactly the beliefs and desires that produced the action. To return to our example above, did the man with the umbrella think it was going to rain? Did he want to look like a City of London banker? Or was he using the umbrella as a cane?

We carry vast numbers of beliefs and desires around in our heads all the time. How do the ones that explain what we do at any one time find their way into our belief and desire boxes? Of all the beliefs in our minds, what makes the belief that umbrellas keep us dry in the rain come into our belief boxes when we want to stay dry in the rain? It's just another absurdly obvious fact about how the mind works, according to the theory of mind, that what makes a belief come into the belief box is its relevance to a particular desire. Or, to be more specific, the beliefs that make their way into the belief box (from "perceptual processes" or memory, by the "inference mechanisms") have some of the same *content* as the desires in the desire box. In our example, for instance, they both *contain* statements *about* rain, and *about* staying dry.

This feature of the theory of mind is so obvious that the figure 3.2 diagram doesn't even make it explicit beyond the four lines inside the belief and desire boxes that are evidently supposed to be statements or sentences that describe the particular desires and beliefs we use to explain or predict behavior. But this component of the theory of mind is its most distinctive and most important feature.

The theory distinguishes between beliefs by their content, the statements they contain, and distinguishes between desires in the same way. What makes the belief that Paris is the capital of France distinct and different from the mistaken belief that Berlin is the capital of France is their *content*: that the former is *about* Paris and the latter is *about* Berlin. What makes the belief that Paris is the capital of France different from the belief that Paris is the

"City of Light" is, again, their *content*: one is *about* where the seat of the French government is, and the other is *about* what makes Paris so romantic. Similarly for desires: what makes the desire to reach the French capital distinct from the desire to reach the German capital is their *content*: one is *about* reaching a French city and one is *about* reaching a German city.

The fact that the relation between these beliefs and desires and countless others is immediately obvious to conscious introspection is crucial to how the theory of mind works. It's the statements that beliefs and desires *contain* that determine which ones get into the belief and desire boxes, which drive the behavior we seek to predict, explain, or both. Without that fact about how they are related, the theory of mind would be completely incapable of explaining anything.

Just think about how any one belief-desire pairing brings about the decisions and actions it explains? For example, what is it that makes the desire to go to the French capital and the belief that Paris is the French capital work together to bring about the action of heading toward the City of Light? The answer, as we've agreed, is obvious, and its obviousness reflects the nature of beliefs and desires: they have *content*—statements that express what they are *about*. The belief is *about* Paris, right? It states that Paris is the capital of France. If the *content* of the belief box had been the mistaken belief that Berlin is the capital of France, that would have caused a quite different set of actions when combined with the desire to go to the capital of France. You would have headed for the city where the Brandenburg Gate is. Same goes for the desire box: it *contained* the statement "I want to go to the French capital." If it had instead contained the statement "I want to go to the German capital" then combined with the belief that Paris is the capital of France, the belief-desire pairing would have explained nothing. But when combined with the mistaken belief that Paris is the capital of Germany, the belief-desire pairing would have explained heading for the City of Light after all.

So it's a crucial component of the theory of mind that beliefs and desires *contain* statements, that they are *about* goings on, people, and things, facts real or imagined, that are relevant to the behavior we use the theory to predict and explain. But how exactly do beliefs and desires have *content*, how is it that they are *about* the world? Are these words we have italicized and will continue to italicize mere metaphors? They can't be if the theory of mind is to do the work it was designed to do. We can begin to get a grip on the way the theory requires beliefs and desires to have *content* and be *about* things by examining how the things we create have *content* and are *about* things.

Consider the red octagonal signs posted at most four-way intersections that lack traffic lights. The red octagon signs are *about* stopping, and they *contain* the command "Stop." Although we call them "signs," they are more than signs; they are "symbols."

What's the difference? It's easy to illustrate: clouds can be a sign of rain, but they're not themselves symbols of rain. Pictures of clouds with parallel lines slanting down from them, however—as in figure 3.3—*are* symbols of rain. But these pictures are not themselves actual clouds, much less actual rain.

Unlike a sign, a symbol *represents* something else. The red octagon sign at an intersection represents the command to stop and the declaration that this intersection is a place where that action is required. That's why and how red octagon signs at intersections are *about* stopping; that's why they contain the statement "Stop." They represent the command to stop where they're placed. Notice, red octagon signs don't represent the English

Figure 3.3
Picture symbolizing rain.

sentence "Stop!" How do we know? Because people who don't speak English all over the world, even in places where English isn't spoken, stop at intersections when they see these red octagon signs. So, in what language do the red octagon signs represent stopping? Well, in all languages or in no particular language. The red octagon signs are *about* stopping, not about the words for it in any language. It takes words in some language (spoken or written) to say what the sign represents, but we shouldn't mistake those words for what the red octagon sign represents. It represents the command to take an action. The stop sign is a symbol, something that stands for something else. It's a representation of something else, what it stands for—an action and a command to perform it where the sign's placed.

The theory of mind works because it treats beliefs and desires as working in something like the same way the red octagon signs work, as representations.

That's how they can be about things and have content. Beliefs and desires have content, are *about* things, because they represent the way symbols do. The difference between beliefs and desires is how they represent. Beliefs represent the way the believer thinks matters are in fact arranged. Desires represent the way the desirer wants matters to be arranged. Beliefs represent in a "mind-to-world" direction of fit insofar as the theory of mind holds that we shape our beliefs to fit reality or the truth. Desires represent in a "world-to-mind" direction of fit insofar as the theory holds that we seek to rearrange reality, the world, enough to make our desires come true.

The way in which desires and beliefs have content and are *about* things, as representations, with these directions of fit, make an indispensable contribution to the explanatory power of the theory of mind. It helps us understand which beliefs and desires make their way into the relevant boxes (though not how they do so), and helps us see how they work together in the "decision-making system" to bring about actions.

All this may not be obvious until we start thinking explicitly about how the theory of mind we all use all the time really works, what its components are, and how they're related to one another. But once we think about these matters, the obviousness emerges. All we're doing is making explicit what we knew implicitly all the time. Another way of putting all these points is that our thoughts—our beliefs and desires—have *meanings* given in the statements they *contain, about* the way the world is or could be arranged.

The obviousness of the theory of mind, now that we have articulated its details, enables us to see that we acquired it in a completely different way from how we acquired any other theory we use: we never spent any time in school learning it, and most likely our parents and siblings never explicitly taught it to us. It's now pretty clear why most of the time when we use it, we aren't even conscious of going through any very complicated process of "inputting" data to the theory in order to draw conclusions—outputs—the way we consciously are of how we use all those theories we learn later in life.

Going back to the passage from Ian Kershaw that explains why Hitler declared war on the United States in December 1941 (Kershaw 2007, pp. 225–226), you'll see that Kershaw is employing the theory of mind pretty much as sketched in the figure 3.2 diagram. And we use the same theory, too, in reading his explanation. After telling us what statements were in Hitler's belief box and what statements were in his desire box, Kershaw leaves it to us to infer smoothly, correctly, and immediately where they came from and how they worked together to explain Hitler's otherwise incomprehensibly stupid action—declaring war on a country that would have had little interest in fighting him, now that it had the Japanese as enemies.

In Hitler's belief box there was this, as a result of the operation of inference mechanisms:

[G]rossly overestimating Japan's war potential, [Hitler] thought its effect [the attack on Pearl Harbor] was far greater than it turned out to be.

In Hitler's desire box, as a result of feelings of pleasure monitored and reported from his body, Hitler could calibrate his desires and identify which he wanted most to fulfill:

[H]e could not have wished for better news than a Japanese assault on the American fleet at anchor. Japan and America at war was exactly what he wanted.

More desires in Hitler's desire box, presumably an emotion the body-monitoring system had been reporting for months:

He had no doubt been itching to do this all autumn.

Another brace of beliefs in Hitler's belief box, as produced from perceptual processes when he was told about the Japanese attack on the United States and as combined with the results of inference:

Now he need hold back no longer. This in itself, he imagined, would turn the battle of the Atlantic Germany's way.

More results of Hitler's inference from other beliefs stored elsewhere in his mind:

Prestige and propaganda considerations dictated that [the declaration of war] should come from Germany, and that he should not passively await a declaration by America.

Still another desire in Hitler's desire box:

To prevent Japan, once in the war, from leaving it at a time that did not suit Germany.

There was also a hope, a combination of an inferred belief in the belief box and a felt desire in the desire box:

The United States, through German intervention, would be forced into a war on two oceans.

Thus does Kershaw explain Hitler's mistake, the practical reasoning to Hitler's decision with the resulting action illustrated in the photo (figure 3.1) at the beginning of this chapter:

From his perspective, therefore, the declaration of war in December 1941 was no great gamble, let alone a puzzling decision.... The decision seemed to him to open up the path to victory which was by the beginning of autumn 1941 beginning to recede. For him, therefore, it was the only decision he could make.

Some of us might well disagree with Kershaw's explanation, but none of us has any trouble understanding it.[2] In chapter 4, we'll explore why it is and how it could be that we've been able, almost from birth, to use the theory of mind. We'll see how it is that, without ever being taught, we employ this complicated theory not just to learn history, but also to take on board just about everything anyone needs to grow up.

2. Here's another example of an explanation of Hitler's surprising actions, this time from an academic historian writing an academic chapter for an academic tome, *War and the Modern World*, volume 4 of *The Cambridge History of War*:

After the Munich Conference, Adolf Hitler regretted having recalled his order to initiate hostilities. His "lesson" from Munich was not to pull back from war again, but to initiate hostilities in 1939. To keep his eastern border quiet while fighting Britain and France he wanted the countries in the east to subordinate their policies to Germany. In the winter of 1938–39, he succeeded with Hungry and Lithuania. Poland's leaders, though they considered substantial concessions in serious negotiations, were unwilling to surrender this country's recently regained independence without a fight. Early in 1939 Hitler therefore decided to attack Poland in the fall and not to become involved in a negotiation that might make it difficult to initiate hostilities, as he believed had happened in 1938. (Weinberg, 2012).

I leave it to my readers to apply the theory of mind to this passage if they haven't already done so.

4 Is the Theory of Mind Wired In?

As we employ it and as our ancestors have used it for at least the last fifty thousand years or so, the theory of mind is complex, detailed, hard to use, and, worst of all, almost impossible to teach. Why impossible to teach? Because teaching almost anything at all and certainly learning anything from someone else already require that teacher and learner both share a theory of mind. That fact makes it hard not to conclude that the theory must be innate, the result of a developmental program encoded by neural genes in our brains. If not, then it must be the case that our brains are strongly predisposed, somehow prepared for or primed by our genetic inheritance to learn, acquire, and internalize the theory on the basis of very early and very limited experience.

Either one of these alternative possibilities explains why we employ the theory of mind without even noticing that we do, why we find it satisfies curiosity better than anything else from earliest childhood, and why we can't give it up, no matter what its weaknesses, failures, and alternatives.

How could so complicated a theory be written in our genes? Or, alternatively, how could we be so predisposed to learn such a complicated theory so soon after birth? If acquiring the theory of mind is a necessary condition for learning anything from someone else—including language—then, hard as it may be to accept, one of these two possibilities must be true.

There is plenty of scientific evidence in favor of each possibility. We'll walk through some of this evidence, from neuroscience, from clinical cases of mental illness, from developmental and cognitive social psychology. We won't be able to determine once and for all whether the theory of mind is completely innate or acquired in early childhood. But we won't have to. Either way, we'll have a full explanation for why we love narratives, effortlessly remember stories, fall for conspiracy theories, and are so sure we understand history.

One way to convince you of the (nearly) innate role of theory of mind in human understanding is to tell a story of how cognitive scientists discovered that it's (close to) hardwired in our brains. A second way is to tell a story about how they discovered that we acquire the theory very early in childhood, as infants even. But the right thing to do is not to tell a story at all. Rather it's to identify the specific regions of the brain where cognitive scientists have found the theory of mind to be "hardwired" and what they've found about how these regions work to control our behavior. We'll do all three, because the stories work, and the science is the right explanation.

Let's start with the second way. Like many stories about babies, my account begins with the twinkling of an eye (Gredebäck and Melinder, 2011).

Subtle as it is, the eye movements of other people secure an infant's eye tracking at a matter of a few weeks of age, as do self-propelled objects.

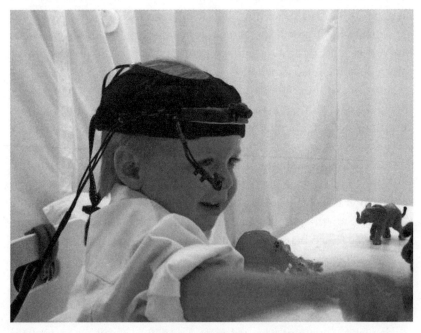

Figure 4.1
Infant with eye motion tracker. http://www.indiana.edu/~dll/images/IMG_1303.jpg

By the age of only a few months, infants are already distinguishing the motion of bodies that looks goal-directed from uniform or random motion. They can do so without any prior exposure to the specific behaviors in question. Expose a four-month-old infant to a human, a robot, or a box moving around a barrier to a location at which each stops. Remove the barrier and have the person, the robot, or the box move along the same path, instead of directly to the location. The infant eye's pupil dilation, its expression of surprise, and dishabituation—amount of time before the child ceases to look—are all greater under the second condition. This response is universally interpreted as showing an extremely early, unlearned interest in and an ability to discriminate means-ends behavior from mere motion, and to be perplexed by the failure of the human or robot or plain box to manifest it. The experimenters describe the result as revealing that "infants already at four months of age are able to interpret other people's actions as rational or irrational." They go on to admit "it is unclear at this stage if the demonstrated early sensitivity to irrational social interactions is innate or based on early experience [within the first three months of life] with rational actions" (Gredeback and Melinder, 2011, p. 4) This may simply be a pardonable overinterpretation, but detecting purposeful behavior is an impressive achievement for any four-month-old. Let's postpone asking how such detection is made—what developmental program lays down the capacity to discriminate purposive behavior from mere motion and how the program discriminates when it gets the chance.

Cognitive scientists have also found, that, at about one year of age, infants begin to make much more sophisticated inferences from observed behavior of others to their unobservable "intentions" (Gredeback and Melinder, 2011). Admittedly, the only basis for this conclusion is the appropriateness of the infants' responses, but they found that, by eighteen months, infants' joint attention on an object cued by an adult's gaze is pretty reliable and that, by two and a half, toddlers demonstrate a grip on "shared intentionality," pushing a ball back and forth when they play with their partners (Tomasello, 2014).

They further determined that, even at sixteen months, infants seem already to have figured out that other people carry around information about their surroundings. When presented with an adult observing where an object was being hidden and with another adult prevented from doing so by being blindfolded, infants met the mistaken guesses by the adult who had not been blindfolded with expressions of surprise and sustained visual attention (Gredeback and Melinder, 2011).

Scientists are quite rightly unwilling to credit young children with having sophisticated cognitive skills until they confirm unambiguously that the children actually have them. Without good evidence, they quite rightly resist crediting a child with anything so sophisticated as a "nested belief"—having the concept of someone else having a belief. For one thing, they can't see into the child's mind to detect beliefs; for another, it's hard to determine exactly what any belief is about, and, for a third, it's even harder to establish that a child has a belief about someone else's belief. Exactly what behavior of the child would reveal this?

It was the philosopher Daniel C. Dennett who first figured out how to answer this question (Dennett, 1978). Thinking about how children respond to a Punch and Judy show (figure 4.2, plate 1) led Dennett to realize the obvious fact staring everyone in the face: if a child believes that someone else has a false belief, then the child has to believe that someone else has a belief in the first place. Punch and Judy shows only work if children have beliefs about the false beliefs of Punch. They laugh when they see him search for Judy in places they know she isn't. Their laughter betrays their belief that he has a false belief.

Following Dennett's insight, cognitive scientists were surprised to discover that, at about four years of age, children have acquired a quite grown-up theory about such false beliefs. By the early 1980s, scientists had set up experiments to detect the age at which children's behavior unambiguously betrays that they think other people have false beliefs (Baron-Cohen et al., 1985). In the standard experiment, something is hidden in the presence of the child, who is required to predict the search behavior of another person. Correct prediction that the person will look in the wrong place requires the child to attribute a false belief to the searcher. For children to grasp that a third person can have beliefs—true and false—about the false beliefs of a second person takes a few more years. But parents know that, from a very early age, children are effectively manipulating us in ways that require us to attribute to them a complicated theory about our beliefs, desires, intentions, and so on. That theory is a pretty full-fledged theory of mind.

Now we can begin to see other complex and interesting components of the theory of mind we all carry around with us, ones we didn't even mention in chapter 3. To begin with, once we have grasped the theory of mind,

Figure 4.2
Punch and Judy show: experimental apparatus for determining the child's acquisition of the theory of mind. https://en.wikipedia.org/wiki/Punch_and_Judy#/media/File:Swanage_Punch_%26_Judy.jpg

there is, in principle, no apparent limit to the number of nested beliefs we can have about another person's belief about still another person's belief and so on *ad infinitum*. The same goes for desires about other people's desires about still other people's desires. Moreover, we can have beliefs about other people's desires, desires about other people's beliefs, and so on in any number of mind-boggling iterations and combinations. However, as Nathan Oesch and Robin Dunbar show (Oesch and Dunbar, 2016), there is an upper limit to the "recursion" of nested beliefs and desires we can keep in our working memory. The limit is about five levels of "A believes that B wants C to believe that D wants E to believe …" Many of these iterations are important in human life. If you think about your willingness to accept paper money, for example, you'll realize it's based on your beliefs about other people's beliefs about still other people's beliefs, and so on. Why else do we accept as valuable small pieces of paper based only on our belief in the "full faith and credit" of the U.S. government? The same goes for all the kinds and combinations of beliefs and desires we have—hopes, fears, likes, dislikes, wishful thoughts: the theory of mind tells us they can be endlessly combined, iterated, and nested inside people's belief and desire boxes.

The achievement of infants and toddlers in learning a theory of mind or deploying an innate one is even more impressive when we appreciate what using the theory requires them to assume. As we've seen, whoever adopts the theory of mind, whether consciously or not, accepts that we all have in our minds "representations" of how the world or some part of it is arranged and the way it could be arranged. Recall our discussion about one of the crucial differences between beliefs and desires, the "direction of fit" difference that is revealed whenever we talk about "wishful thinking."

The crucial role of this component of the theory of mind is reflected in the way it gets combined with the nesting of belief and desires that is required for the child's acquisition of language. A toddler's achievement of realizing that the sounds its parents make are meaningful speech and not just sounds requires that the toddler have beliefs about what its sound-making parents want and believe, and, in particular, what those beliefs and wants are *about*, what their *direction of fit* to the world is. The child has to believe that its parents have beliefs and desires that represent the way things are and the way the parents want things to be. Until it has nested beliefs about the content of its parents' beliefs and desires, the child is treating the sounds its parents make the way we treat the sounds a squeaky wheel makes, trying anything we can think of to get it to stop, without thinking that the wheel's actually telling us something about what it wants and thinks. We'll come back to this relation between the theory of mind and language again in later chapters.

So, before children have acquired language and almost anything else we can teach them, they have taken on board a pretty hefty psychological theory. How could they have done it?

Here's why some cognitive scientists argue that the theory of mind is "innate," hardwired, laid down in the developing brain by a genetically encoded process. The theory of mind is universal in us humans, present in all cultures. If acquired at all, it's triggered or learned quickly, very early, in infancy, and under a vast range of circumstances. How could we all learn the same theory so quickly, so uniformly, and so accurately? Maybe, as children, we don't learn it at all. Maybe we just have it from birth, requiring a little priming in early infancy to trigger its expression. Even children blind, deaf, or both from birth seem to acquire the theory easily and early.

Only a very little learning of no special kind seems required—if any at all. Think of blind, deaf, and mute Helen Keller, who somehow has acquired the theory of mind, as she is portrayed in the 1959 play and 1962 film *The Miracle Worker*.

Moreover, children employ the theory of mind automatically, and, indeed, it's hard for them to stop from invoking the theory to explain and predict, even when it doesn't apply. They overshoot, as we all do even as adults, seeking theory-of-mind explanations for things that happen for no reason at all.

Motivated by these sorts of considerations, cognitive scientists in clinical settings began to uncover a much more powerful reason for treating the theory of mind as either innate or triggered by only the slightest provocation in early childhood experience: the behavioral pattern labeled "autism."

"Autism" or "autism spectrum disorder" are fraught terms, subject to controversies at the intersection of science and politics. In recent years, autism has become a touchstone for the politics of difference, diversity, and demands for the treatment of unusual human traits as equally valuable ways to function, each with its own contribution to humanity. To explore what autism reveals about the theory of mind, we have to negotiate a minefield of passionate disputes at the intersection of science and human values.

Autism was first diagnosed variously, as an "illness," a "disease," a "disorder," or a "defect." It's worth noting, to begin with, that most physical diseases or illnesses are known to have a number of symptoms but only a single cause—that's what makes each of them a distinct disease or illness. That is not yet true of most mental illnesses or disorders. Our taxonomy of mental illnesses, as set forth in the *Diagnostic and Statistical Manual of Mental Disorders*, 5th edition (*DSM-5*) and its predecessors, is a classification in terms of symptoms—effects, not causes. Since the same effects can and often do have different causes, the various editions of the *DSM* don't actually identify kinds of mental illnesses with much precision. And some effects that look pathological to one physician, culture, or epoch may seem quite unpathological to another. Consider homosexuality, once classified as a "disorder" in the *DSM*, but no longer.

By contrast with other mental conditions, autism spectrum disorders were initially thought to have a fairly narrow range of causal agents; their onset was attributed to the failure of one or another component of the child's acquisition and employment of a theory of mind.

Thirty years ago, social psychologists established that children with autism fail the false-belief test at ages when normal children and otherwise mentally disabled children (for example, those with Down's syndrome) pass the test. When the genetic basis of autism was widely recognized and accepted in the 1980s (see Baron-Cohen et al., 1985), biomedical researchers sought to identify the defect in the brain resulting from the genetic defect or defects associated with autism. The next step would be to locate the defect in the DNA.

To say that autism has a genetic basis is not say that is transmitted from parents to offspring. Indeed, researchers have found little indication that the condition is genetically hereditary. But it's well known that when one identical twin is autistic, the probability the other twin is as well is higher than 50 percent, and the probability that the other twin has some related developmental "disorder" is higher still. This suggests that autism probably has some early, possibly prenatal cause that is carried into both twins by DNA copying. There is, however, almost certainly no single gene for autism. This should be no surprise given what we already know about the quantitative genetics of most human traits that have been studied. Even height, the most strongly inherited human trait is correlated with literally scores of genes no one of which seems to contribute more that 2 percent to the probability that an individual attains a certain height or not.

Autism, then, is a genetic trait in the sense not of inherited but of *somatic* genes, the genes in the body's cells that control what the cells do. In particular, it's caused by variations in the genes of the cells that build certain parts of the brain. Genetic variants can be inherited from parents if they are present in the sperm or ovum. But they can also be the result of "breakdowns" in the copying of somatic genes soon after fertilization that produces the zygote's full complement of twenty-three chromosome pairs. Then they won't be inherited, even though they are genetic in origin. Twins start out sharing an almost perfectly identical sequence of 3 billion nucleotides in their genes, which then get copied over and over through development in all their cells. The copying produces a relatively small number of inconsequential differences—copy "errors"—in their DNA sequences. (The quotation marks around "breakdowns" and "errors" remind us that all genetic variations have their source in the always imperfect process of gene copying, and that whether the resulting trait is favorable or harmful depends entirely on the environment it interacts with.) Since the chances these copy errors will be the same in two twins are vanishingly small, the variations in autistic twins' genes that produce the same symptoms almost certainly happened before the original division of the twins' zygote. And because autism doesn't have a clear pattern of inheritance across the generations, it's also probably the result of combining two genomes—the mother's and the father's— neither of which, alone, has enough genetic copy "errors" to result in autism, but which together do have enough (Brandler and Sebat, 2015).

But which genes are involved? Some 60 percent of all genes are expressed in the brain, that is, make proteins that build and maintain the brain, and different amounts or assemblages of protein structures and different orders of activity can produce much the same developmental outcome or behavior. Autism is therefore the result of a variation somewhere in the building of the brain or in some regions of it that involves a relatively rare combination of a large number of genes, most of which are common to everyone, but some of which are rare. The particular combination will almost certainly differ from case to case, even as it produces the "same" result, namely, the range of behavior that is symptomatic of autism. Although we still have much more to learn about the condition, what the evidence shows so far is that autism is a result of differences in the genetically encoded program of brain development.

Is the failure of children with autism to pass the false-belief test also a failure to fully deploy the theory of mind? Influential researchers like Simon Baron-Cohen (Baron-Cohen, Leslie, and Frith, 1985) and many others thought so in the 1980s. They inferred that the theory of mind must itself be largely the result of a genetically encoded, hardwired program of neurological development. The wiggle word "largely" acknowledges that, almost certainly, some crucial, though as yet undetected, triggering stimuli in earliest childhood, most likely in infancy, were also required for the child to begin fully deploying the theory of mind. Researchers hypothesized that in autistic children, when the crucial triggering stimuli occurred, the theory of mind fails to deploy.

This view of the nature of autism led cognitive scientists to seek a domain or module in the brain that specifically and narrowly "subserves," "realizes," "implements," or "instantiates" the theory of mind. Most of their research has employed functional magnetic resonance imaging (fMRI), a technique for inferring brain activity from the differential uptake of oxygen during cognitive tasks of various sorts (and one whose well-understood limitations will eventually give way to finer discriminations employing more direct measurements). Another approach to localization employed by cognitive neuroscientists scientists is transcranial magnetic stimulation (TMS), which focuses a magnetic field on a small area of the brain, where it interferes with the neurons' electrical polarization up to 6 centimeters below the scalp. Whereas fMRI identifies regions of activity by their use of oxygen, TMS identifies them by temporarily disabling the effects of their neurons on behavior

Every month, research employing fMRI and TMS by a large number of cognitive neuroscientists reports new findings about the localization of cognitive activity and emotional response. The regions identified are relatively large and almost certainly responsible for a range of distinct, though related, behaviors. Moreover, several of the regions work together in bringing about any one activity, and each is involved in more than one identifiable behavioral task. In fact, the main challenge in this research is designing behavioral tasks that are so constrained, specialized, and measurable that they are produced only by a small number of these distinct regions (Gweon and Saxe, 2013; Saxe, Carey, and Canwisherm, 2004).

Identifying brain regions with distinct functions—modules—is an iterative process: researchers start out with localization in relatively large areas due to co-occurrence of behavioral deficits with visible abnormalities—lesions, congenital deformities, stroke sites, and so on. They then correlate these deficits with findings from fMRI and TMS in normal brains and devise behavioral tests that activate specific regions. Subsequently, they try to design new behavioral tests that elicit activity in smaller and smaller parts of these regions. The limitation here is, of course, the researchers' ability to design such tests and ensure that all subjects understand instructions and interpret stimuli in a closely similar way. Testing for the localization of theory of mind requires that subjects all share the same theory of mind and that they want to aid the researchers by using the theory while under fMRI scanning. The researchers' tasks are to design experiments in which subjects have to attribute false beliefs to others, either upon listening to stories or viewing cartoons, films, or demonstrations of behavior inexplicable unless the agents are acting on false beliefs.

Using both fMRI and TMS, Tobias Schuwerk, Bertold Langguth, and Monika Sommer identified five regions of the brain as loci that work together to deploy a theory of mind (figure 4.3): (1) the left and right temporoparietal

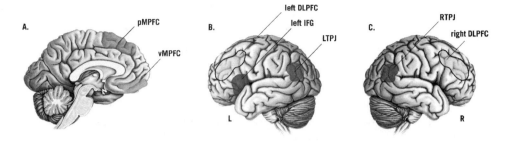

Figure 4.3
Five regions of brain identified by both fMRI and TMS as loci that work together to deploy a theory of mind: the left and right temporoparietal junction (LTPJ and RTPJ); the left and right dorsolateral prefrontal cortex (DLPFC); the left inferior frontal gyrus (IFG); the ventral medial prefrontal cortex (vMPFC), and the posterior medial prefrontal cortex (pMPFC). From Schuwerk, Langguth, and Sommer, 2014, fig. 1. Courtesy of *Frontiers in Psychology*.

junction (LTPJ and RTPJ); (2) the left and right dorsolateral prefrontal cortex (DLPFC); (3) the left inferior frontal gyrus (IFG); (4) the ventral medial prefrontal cortex (vMPFC); and (5) the posterior medial prefrontal cortex (pMPFC) (Schuwerk, Langguth, and Sommer, 2014).

The localization of theory of mind to this small number of particular brain regions is one of the most robust and well-replicated results in neuroimaging (see Saxe et al., 2004). Using different tests and stimuli over a range of different male and female subjects of different ages, speaking different languages, and from different cultures, researchers showed that the same five regions of the brain were specifically activated by tasks requiring theory of mind.

Transcranial magnetic stimulation has the advantage over fMRI in that it can be employed to temporarily inhibit normal functioning in specific regions of the brain, especially ones close to its outside surface and near the skull. In effect, TMS can switch on and off regions of the brain by interfering with or stimulating patterns of electrochemical discharge. TMS studies naturally offer the prospect of a finer-grained geography of specialization among brain regions than fMRI can provide. They have strongly confirmed results from fMRI about the specific regions of the brain that are differentially involved in theory-of-mind reasoning. In fact, the TMS studies have begun to help identify the specific contribution of each of them to the mind-reading task (Kalb et al., 2009). Of the five regions highlighted in figure 4.3, transcranial magnetic stimulation has begun to suggest that one specific function of the right temporal medial junction (RTPJ) is to enable the experimental subject to adopt the perspective of the other person (agent) the subject is thinking about and whose behavior the subject is to explain or predict (O'Connell et al., 2015). By contrast, TMS studies suggest that the posterior medial prefrontal cortex (pMPFC) does the opposite, enabling the experimental subject to distinguish his or her own perspective from that of the other person (agent) whose thoughts the subject is tracking, an essential task in false-belief reasoning (e.g., Schuwerk, Langguth, and Sommer, 2014).

Taken together, the research provides evidence that the theory of mind is localized to a small number of specific brain regions acting together as a mental "module." Besides its localization, the imaging research reveals that it's "domain specific": the regions of the brain that encode and deploy theory of mind are not active in other tasks that seem quite similar to the theory's proprietary activities. For example, when the brain detects other kinds of "falsehoods" besides false beliefs, for example, directional signs pointing in ways the experimental subjects know are wrong, or pictures that are known by the subjects to be inaccurate, the theory-of-mind regions are inactive. When subjects are told stories or shown cartoons that don't involve false beliefs, the pattern of nonactivation of these regions and activation of non-theory-of-mind regions is quite distinctive. The theory-of-mind regions are also distinguishable on fMRI from other nearby regions involved in cognition that is focused on human purposeful behavior that does not involve either false or true beliefs or desires.

The theory-of-mind regions have several other features characteristic of special-purpose mental modules deployed in specific regions of the brain. As we've seen, theory-of-mind capacities develop in a fixed pattern in children, and they break down in characteristic ways that are reflected in medical diagnoses. We know from our own experience that these capacities operate quickly, often unconsciously, and automatically to explain or predict other people's behavior in ways that we're often not even conscious of.

About the strongest evidence against this idea that the right temporal parietal junction is itself a theory-of-mind module is the fact that, as fMRI research has shown, it is also active in some presumably non-theory-of-mind tasks, especially attention tasks (O'Connell et al., 2015). It may turn out, of course, that finer-grained fMRI investigation or the use of other, more refined techniques may differentiate structures in the right temporo-parietal junction that deliver each of these activities independently.

Functional magnetic resonance imaging has also enabled neuroscientists to locate the regions of the brain that "subserve" a different kind of understanding, the kind involved in identifying causes in processes that don't involve purposeful behavior or intentional action, and that don't employ a theory of mind at all. These regions of the brain (the frontopolar cortex, the dorsolateral prefrontal cortex, and the motor cortex) are involved in mathematical processing, logical and spatial reasoning, and visual modeling. They are all well away from those involved in theory-of-mind reasoning, indeed, all the way on the other side of the brain. It's interesting that mathematically gifted male adolescents show patterns of brain activation in three regions different from those active in average-ability students dealing with the same problems. These mathematically gifted kids are often less gifted in their employment of the theory of mind (Yun et al., 2011).

But the findings from neuroscience are not just that brain regions presumably subserving natural understanding are distinct and distant from the regions that subserve the theory of mind. These two sets of regions have also been shown to inhibit, suppress, and obstruct each other's activity. The regions of the brain that subserve theory-of-mind cognition are components of a larger domain—the default mode network (DMN)—whereas the regions involved in mathematical reasoning and visual modeling are part of another larger domain—the task positive network (TPN). The DMN drives social information processing, whereas the TPN subserves reasoning about physical objects (Raichle et al., 2001). What's more, fMRI studies have revealed that these two networks appear to be antagonistic: when either is active, the other is suppressed. There is a competition between theory-of-mind reasoning and physical-object reasoning in the brain. Behavioral experiments reveal that mathematical tasks interfere with responses that require theory-of-mind reasoning. Just as we might suspect, many autistic spectrum disorders are associated with higher levels of "fluid intelligence" and visualization ability, the kind of ability that makes for high achievement in math, but that correlates with lower levels of social cognition (Simard et al., 2015). The antagonistic relationship between these two domains helps explain why most of us, with fully functioning theories of mind, prefer to learn about science through narratives and why this doesn't work very well.

The competition between narrative and analytical modes of thinking is also reflected in the enhanced visualization powers associated with reduced social functioning in some dementias. The opposite also occurs: heightened theory-of-mind capacities are often associated with mathematical deficits (for a review of some of these studies see Jack, Connelly, and Morris, 2012).

In short, neuroscience has begun to amass a good deal of evidence that the theory of mind is somehow inscribed in highly specialized regions of the brain, ones that initially showed deficits in people with autism. It has shown that in neurotypical people, transient interference with this brain region temporarily disturbs theory of mind reasoning. It has provided evidence that the brain's employment of this theory is antagonistic to its employment of reasoning required to understand nonpurposive, physical or mechanical behavior, and that there are differences in the brains of people better than average at reasoning about nonpurposive behavior and worse at theory-of-mind reasoning.

It was at this point that autism researchers dropped out of the research program to locate the theory of mind in a specialized module of the brain. They did so for several reasons. For one, they began to discover that there are other, separately identifiable dimensions of cognitive, motivational, and emotional activity where autistic individuals differ from "normal" or "neurotypical" ones. For another, these researchers turned their attention to the ability of autistic people to deploy cognitive strategies other than theory of mind to cope with social interaction, including the prediction of other people's behavior. And for a third, the achievements of autistic people like Temple Grandin led them to challenge the taxonomy of psychopathology, at least for some people diagnosed with the condition. Autism rights advocates have increasingly argued that, like homosexuality, autism does not require a cure but should be accepted instead as an alternative set of behaviors that may in fact facilitate some creative achievements.

Meanwhile, independent of any simple association with autism, the localization of the theory of mind to a specific domain or module of the brain is well established. But how could such a complicated theory be laid down in our brains by our genes or DNA? After all, just getting the brain built in the first place seems a formidable undertaking for a bunch of macromolecules. Surely, the theory of mind is not simply written down in the alphabet of the four nucleotides of DNA molecules, even if they're three billion base pairs long, but, rather, is something we as infants, babies, toddlers, or young children learn from experience as we grow up. That view drives a lot of the resistance to the notion that our cognitive abilities are hardwired.

And yet there is clear evidence of our acquiring these complex cognitive abilities very soon after birth on the basis of so little experience that there must be some innate "device" or hard-wired "module"—or some detector of early stimuli to trigger our acquisition and deployment of the theory of mind immediately on encountering them. How might such a triggering mechanism work?

Imprinting by ducklings on their mother or on any moving object of roughly her height in the right time window of early development is the oldest well-known example of early experiential triggering (Lorenz, 1949). Another is the Nobel Prize–winning discovery of David Hubel and Torsten Wiesel that kittens' normal visual abilities depend on having a certain experience that organizes identifiable parts of their brains during a brief critical period just after birth. It's well established that, without any prior experience, infants respond to the presence of snakes, and both they and adults detect spiders and snakes more accurately and more quickly than other stimuli, even without prior experience of either. And this innate snake/spider detection ability can be rapidly converted to a fear/flight response with minimal learning. Our widespread antipathy of snakes and spiders, though not innate, is at least close to innate (Öhman and Mineka, 2001).

Noam Chomsky famously noted that every normal prelinguistic infant can learn any language at all, even from highly imperfect speakers, after only a few months of exposure. We acquire a capacity to decode and encode an infinite number of spoken sentences within a brief period of time after hearing even a small sample of expressions, however ungrammatical or otherwise defective they might be. All this suggested to Chomsky that we humans are born with a cognitive device that enables us to construct, discover, extract a set of syntactical rules for the language we're exposed to, and begin using it early in childhood. Contrary to popularizations of his work, Chomsky believes that this device evolved as the result of selective pressure not for language, but for a more basic role in cognition and reasoning (Hauser, Chomsky and Fitch, 2002). Perhaps, in order to produce language, our ability to think syntactically needed to be harnessed with a theory of mind (a subject we'll return to in chapter 11).

All of which is to say there is much evidence to support the conclusion that a theory of mind is a fundamental and universal component of our cognitive equipment, largely innate, requiring only the slightest of triggering by stimuli in early childhood and emerging at the same time among most of us, regardless of our circumstances. This component can be damaged or disabled in the developmental process that builds our infant brains, and, like a distinct organ, it is localized to particular parts of those brains. Moreover, the ability of our brains to use the theory can be turned off by experimental intervention as TMS studies show.

The conclusion that the theory of mind is a (nearly) innate component of our cognitive tool kit and that it is indispensable for learning language and, indeed, for learning anything at all from other people raises the deep and yet also obvious question of how such a trait evolved. If it turns out to be an indispensable prerequisite for life as we humans know it, the theory of mind almost certainly must have an evolutionary pedigree.

Although we'll address its pedigree in greater detail in chapter 5, a few remarks are worth making here. Evolutionary anthropologists have discovered that the theory of mind and the more fundamental abilities from which it emerged played a central role in our very survival as a species (Tomasello, 2014; Hrdy, 2009). Their findings vindicate what fMRI studies have shown about the theory of mind being a distinct cognitive module and reveal how so complicated a cognitive capacity can get itself (practically) written into our genomes. In short, these findings establish that the theory of mind couldn't work at all unless it was almost completely hardwired in the developmental program of our brains.

But the Darwinian pedigree of the theory of mind also raises serious questions about the theory's explanatory power—questions whose answers end up completely discrediting historians' claims to provide real understanding of human affairs.

5 The Natural History of Historians

The strongest evidence suggests that our species, *Homo sapiens*, emerged in Africa some 300,000 years ago, quite recently in the larger scheme of things (Hublin et al., 2017). If we're to last as long as the dinosaurs, we'll have to be around for another 250 million years or so. Along with several other contemporaneous species in the hominin line—Neanderthals, Denisovans, *Homo floresiensis*—*Homo sapiens* evolved from primates who had moved out of the deep African forests and onto the margins between the wooded and grassland savanna about two million years ago. Why they did so is not entirely clear. It might have been in response to climate change, to competition with other species, or, more likely, to a combination of both. Africa was becoming more and more arid, and the forests were shrinking. Other primate species still in the forests, such as gorillas, chimpanzees, and bonobos, were probably better adapted to them and pushed out the weaker, less adapted species, our ancestors. Or it might have been that some hominins simply followed the tree-lined riverbeds out into the savanna, where natural selection adapted them to survive—but only just.

In any case, long before we came along, the primate lineage out of which we and the other species of the genus *Homo* evolved, faced a huge problem or, alternatively, a huge opportunity. The problem was almost immediate extinction; the opportunity was potential domination of the entire food chain. But the problem and the opportunity were really two sides of the same coin. You could think of that coin as a glass half full or half empty—or as a "Necker cube" drawing (figure 5.1), in which the bottom square can look like the front or the back of the cube, depending on how you view it.

Solve the problem and dominate; fail to solve it and become extinct. Only the immediate ancestors of our species eventually solved the problem

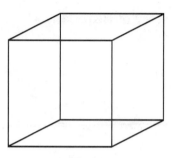

Figure 5.1
Necker cube.

and made the most of the opportunity; the ancestors of other species did not. We *Homo sapiens* are the only one of the four known species of *Homo* to survive. Although the evidence from population biology is, to say the least, scanty, continually facing extinction for most of the two million years that they were evolving on the savanna, our hominin ancestors appear to have found themselves trapped in a genetic/population bottleneck. With their small numbers, the total gene pool was also small, and therefore the source of new variations that might have solved the problem and increased their numbers was also very small (Stringer, 2013).

Let's first consider the problem/opportunity facing hominins from the problem perspective. Early hominins were at the bottom of the savanna food chain, an ecosystem dominated by large animals, top predators, who were faster, stronger, and better armed—and who greatly outnumbered them. Although the hominins' diet consisted mainly of plants—grasses, seeds, and tubers—eventually, they began to scavenge meat from carcasses brought down by the large predators.

The threat from these predators was combined with three other problems: hominin females were having too many offspring, the birth interval between offspring was too short, and the period of childhood dependence too long. All that meant too many children unable to fend for themselves, no opportunity for adult females to provide for themselves, much less to help males provide for all those children they had to carry around for years, tending to them even longer. Because they'd be too large when fully developed for the skulls around them to make it through hominin mothers' birth canals, the brains of hominin children, unlike those of other primate children, had to grow mainly after birth, and this made hominin children dependent on their elders for the better part of a decade (Falk et al., 2012). Hominin mothers were having children too close together because they weren't breast-feeding as long as other primate mothers, and breast-feeding shuts down ovulation. What's more, the larger number of children needing to be fed reduced the average amount of nutrition each child could be given.

How could our hominin ancestors solve these problems? Was there a way to turn some of these problems into opportunities to solve others? Could all three of these problems somehow turn an extinction threat into an evolutionary opportunity? Well, the combination of a long period of dependency and a large, growing child's brain could be exploited: dependent offspring are compliant and, with their large but maturing brains, they could be made to learn a great deal during the period of compliance and dependence, if only they could be taught by adults. Technologies like tool making, hunting, and fire keeping could be transmitted, instead of dying out with their inventors. But teaching required learners to have the ability to imitate the behavior of their teachers, of course, and to guess what they wanted to teach. It required teachers to figure out how to motivate their pupils not only to learn and but also to learn the right things. And it required both parties to accurately predict each other's behavior.

If there were too many children for one mother to handle, perhaps the burden could have been shared among several mothers or grandmothers, freeing a large number to gather or scavenge while a smaller number tended to and protected the children. But this, too, required our hominin ancestors both to reliably predict the behavior of others and to enter into and abide by agreements, in short, to cooperate in child rearing.

But mothers who left their children to be cared for by others couldn't nurse them, and, deprived of protein, their children's brains wouldn't develop, and the children wouldn't survive, or, even if they did, they wouldn't be able to learn what they needed to. There had to be another reliable source of protein besides mother's milk. If the adult males could team up to hunt large prey animals themselves instead of just following predators around hoping to find the prey animal scraps they left behind, that might solve this problem.

Achieving any, let alone all, of these solutions, however, required a theory of mind. To get a better sense of the benefits of acquiring one, let's consider the problem/opportunity facing our ancestors from the opportunity perspective. Out on the veld, some hominins were able to expand their diet from grasses, seeds, and tubers to one that included increased protein in the form of meat by scavenging. Reduced consumption of plant foods, which were harder to digest, and increased consumption of meat, which was much easier to digest, especially when cooked, reduced the need for large gut size (and freed up calories to build other parts of the body). Increased protein consumption led to larger brain size, along with much smaller teeth. It's the shape, size, and mineral composition of our ancestors' teeth and, increasingly, their preserved DNA that reveal this sequence in detail (Teaford et al., 2000). And the long periods of childhood dependence that resulted from the need for the children's larger brains to develop after birth provided opportunities for learning that their larger brains could take advantage of. Increased intelligence could then combine with collaborative child rearing, a propensity shared with only a few other, unrelated species—tamarin monkeys, elephants, canines—but not with other primates (Hrdy, 2009).

The initial division of labor between females who cared for infants and those who gathered plant food could significantly enhance productivity among the hominins, but it also presented a serious problem of trust. How could the gathering mothers trust the caregiving ones to tend to their children? And how could the caregivers trust the gatherers to provide for them and for their children? Who could be trusted? It would take a theory of mind to figure that out.

Among adult male hominins, there were several significant productivity-enhancing opportunities on the savanna. First, there was the opportunity to share the benefits from experience and particularly from specializing in useful technologies like tool making, especially if these could be reliably transmitted to the next generation. Second, there was the opportunity to significantly increase the hominins' intake of protein by shifting from individual scavenging of animal carcasses left by large predators to cooperative hunting of large prey animals, provided, of course, the hominins could learn to gang up on the megafauna.

Any hominin species that could capitalize on these opportunities would end up having more children, and more of these children would survive. When at last they did capitalize on them, our ancestors were able to breakout of the long genetic/population bottleneck they found themselves in. But it took them a million years or so to make the most of these opportunities (Powell et al., 2009). What were they waiting for?

A theory of mind.

The critical problem early hominins faced was that of cooperation. Even hominin scavengers would have faced what amounted to the prisoner's dilemma at the earliest opportunity to cooperate rather than go it alone. Imagine you're one of two male hominin scavengers alive at that time. Having spotted a carcass worth scavenging, you and your fellow hominin both can anticipate that the large predator responsible for it will shortly return to finish feeding on its kill. One of you needs to stand guard while the other secures the remaining meat. If it's you, the other scavenger will get it all and you'll get none, and vice versa. But if neither of you stands guard and you both gorge on the carcass at the same time, there's a good chance you won't notice the predator when it circles back and adds you both to its kill. Because, as it's framed here, there is no solution to the dilemma, the luckless hominins who faced it usually died off.

Evolutionary game theorists recognize, however, that we might not be in a simple one-time prisoner's dilemma. We might be "playing" a different game in which there are a number of cooperative solutions beneficial to both players (Axelrod, 1984; Henrich et al., 2004). Returning to you and your fellow hominin scavenger, if you find yourselves repeatedly in the same situation, you have several better options than getting killed or passing up the scavenging opportunity altogether. Perhaps the best of these involves the well-known "tit-for-tat" strategy. You can take turns guarding a little and scavenging a little the first time you're out scavenging together, and if the other cheats, you won't scavenge with him again. But if he doesn't, you don't cheat either, and you both do the same next time you scavenge together. Under most realistic circumstances, you'll both do as well with the tit-for-tat strategy as with any other, and far better than just trying to gorge yourselves while risking the undetected return of the predator when it comes back to finish off its kill.

Game theorists have identified plenty of other strategic interaction situations most likely faced by hominins on the savanna (Henrich et al., 2004). In a "cut the cake" situation, as a hominin hunter, you bring in a whole carcass, and have to decide how much to give your family and how much to give another hunter's family, while the other family has to decide whether to accept your offer of at least some meat or perhaps fight over the whole carcass, at potentially deadly cost to both families. Offer too little, and everyone loses. Offer too much and you deprive your family. What's the best strategy? Well, game theorists have shown that, in realistic cases where this situation repeats itself over and over, the best long-term strategy for both scavengers would be to offer half of the carcass to the other family every time (Skyrms, 1996).

Then there is the "stag hunt" situation. You and your fellow hominin hunter are lying in wait to trap an antelope when you see a meerkat not ten feet away. If you go for the meerkat, you'll almost certainly be able to kill it, but your sudden movement will spook the antelope. What to do? The antelope provides much more meat, but there's a good chance it may not even wander into your trap. Game theorists would tell you that, if you encounter this situation only once, your best strategy is to go for the meer-kat. But if, however, you find yourself in the same situation over and over again with the same partner, they would say that your best strategy would be to forgo the short-term gain of the meerkat and wait for the antelope to wander into your trap (Skyrms, 2003).

The "ultimatum game" describes another, more complicated situation. You and your fellow hominin are out hunting together. You're better at hunting gazelle, and he's better at hunting springbok. Before you go out, you two need to agree on how to share the kill, if any. He proposes that, if it's a springbok, he'll take three-fourths and you'll take one-fourth of the carcass since you wouldn't have been able to kill the springbok at all, but if it's a gazelle, you'll split the carcass fifty-fifty. Do you take the deal, even though he'd never be able to kill a gazelle without you? No. You propose instead that either you get three-fourths of the carcass if it's a gazelle or you share whatever you two kill, whether gazelle or springbok, fifty-fifty. Although there are lots of other ways to split the kill, without some agree-ment, you two and your families will get nothing. The question is, what's the best deal for everyone? Again, it turns out, according to game theorists, that, in most circumstances, the "fair" split—fifty-fifty—is the best deal for hominin hunters interested only in their own survival, and very likely with no interest in "fairness" (Henrich et al., 2004).[1]

Where does the theory of mind come into these strategic interaction situations faced by hominins on the veld? Well, in the case of nonhominin

1. Here it's important to bear in mind that any single individual strategic interaction situation two beings are likely to face can be embedded in a large number of different "multiplay-multiplayer games" and that calling the interaction situations "games" only means that their outcomes depend on what strategies their players employ.

hunting animals, natural selection didn't need to select for a theory of mind among them so they could find their way to optimal strategies in repeated prisoner's dilemma, cut the cake, and even stag hunt situations. All it had to do was filter out hardwired strategies until the only ones left in a nonhominin species of hunting animal were tit-for-tat or other cooperative strategies. But it's obvious that hominin interactions, even when they had simple payoffs in single interactions, were really far too complicated for the best strategy to be written in the hominin genome. Hominins who expected to survive and prosper in the presence of other hominins they could compete against or cooperate with, were going to need a pretty powerful theory of mind. In most cases, it would have to be complicated enough for them to think through a problem like "What should I do if I want that stranger to believe I want this instead of that?"

But long before hominins had to face strangers and figure out their beliefs and desires, they were going to deal with the same problem around their own campfires. And they would have to solve it to have any chance of surviving long enough to even meet up with unrelated hominins.

But even before prisoner's dilemma and other competitive situations faced by interacting strangers emerged, there were plenty of opportunities for hominins to cooperate and coordinate with their relatives—partners, parents, children, and even an occasional cousin or nephew (Hrdy, 2009). Natural selection for survival of the fittest strongly selects for cooperation among close kin. So long as what you do for your kin results in more offspring carrying your genes, selection will favor conferring unreciprocated benefits on them—sharing your food with them (rather than free riding on their efforts at scavenging, hunting, gathering), teaching everyone in the family your hunting and tool-making secrets and skills, and so on (Sterelny, 2012).

Knowing how to make effective tools was particularly perishable in our hominin ancestors' time; indeed, it was all too likely to perish with its knowers. To begin with, the knowledge had to be reliably transmitted from generation to generation in large enough populations in order to last long enough to be perfected. It took a million years or so to get from the hand ax to the compound ax, one with a handle. The main reason was most likely the long genetic/population bottleneck in hominin evolution. The compound ax was almost certainly invented many times. It doesn't take rocket science to bind a wooden handle onto a piece of stone. But almost all of the very small kin groups that hit upon this innovation died out before the invention could spread. So it had to be reinvented over and over. This made cumulative improvement and refinement of tools extremely difficult, if not impossible. Indeed, the difficulty of reliably transmitting how to make productivity-enhancing tools from one generation to the next and from one small group to the next in the starting populations of hominins is in part what kept their numbers low. It's what gave rise to the long genetic/ population bottleneck they came to be trapped in (Powell et al., 2009). Now, consider what would have been needed to convey the design of, first, the hand ax and, then, the compound ax, from generation to generation in numbers large enough to ensure reliable transmission of the technology. The young members of a kin group having this technology would need to be taught and be able to learn. But teaching and learning are themselves cooperative practices, involving coordination and close imitation. And that requires what cognitive scientists call "shared" or "joint intentionality."

The ability to closely imitate is one that other primates don't share with humans or with our hominin ancestors. Other primates are smart. They can emulate the results of actions. If they see you use a tool to snag some food, they may come to use the tool in similar ways. But they don't have whatever cognitive capacity it takes to actually imitate complex behaviors or actions. They can't even tell when you're trying to get them to imitate your movements and actions (Tomasello, 2014).

As a matter of fact, acquiring the ability to teach and to learn is arguably more complicated than acquiring the technology of building compound axes. If teaching and learning that technology also required first learning and teaching the more complex theory of mind, it's no surprise hominins were around for a million years before inventing the compound ax. Accomplishing the simpler task—building a compound ax—required accomplishing the harder task—learning how to imitate and figuring out how to teach. And because teaching required both parties to apply the theory of mind repeatedly, over and over again, each party would have had to frame beliefs about the beliefs of the other party, beliefs about the desires of the other party, desires about that party's beliefs as well as desires, and sometimes each of these, over and over again.

If everything we ever learn from others requires that we already have the ability to learn, there is only about one solution to the problem of how we ever accomplished the original feat, acquiring the theory of mind, or at least as much of it as is needed for learning by imitation—the theory had to have been either innate or, once learned, faithfully preserved from generation to generation by teaching. Because, without it, the evolutionary path to us humans could never have gotten started. And since the path not only got started but succeeded in producing us, it's no surprise that at least much of the theory of mind is hardwired or as close to hardwired as any cognitive capacity can be. Had the entire theory needed to be taught and learned, the chances are we humans would never have evolved, much less survived and prospered.

But then the question arises, how was all this possible? Just because having a reliably transmitted theory of mind turns out to be necessary for hominin survival and required for producing *Homo sapiens*, there is no guarantee that the theory would in fact emerge. Do we have any idea how that might have happened?

Actually, we do.

The theory of mind emerged from a more basic ability we'll call "mind reading." The label is imperfect since lots of animals with the ability to mind read don't have any idea of what minds are, or even whether minds exist to be read in the first place. For our purposes, "mind reading" is the ability to successfully predict the behaviors of other animals. It's an ability that emerged long ago in mammalian evolution, and perhaps even before that, in vertebrate evolution. The ability both to track prey and to avoid predators relies on it. Think about how a hawk plots an optimal trajectory to intercept its flying prey or even how a chameleon changes color to camouflage itself in the presence of a predator. Both engage in behavior that is environmentally appropriate, that will enhance their survival and increase their chances of reproduction. Natural selection long ago began selecting for a hardwired ability to track predators- and prey, as well as to predict the behavior of possible mates, conspecific mating competitors, and actual offspring. Hawks don't have to learn how to use their hardwired mind-reading ability; they simply grow into it.

By the time elephants, dolphins, canines, and primates emerged, mind reading, was a well developed ability, a bit of "know-how." But it was becoming even more sophisticated and nuanced and was also beginning to be fine-tuned by learning (Lurz, 2011). And from this point onward, at least in the primate line, natural selection eventually shaped the mind-reading ability into a theory of mind that was almost completely hardwired and available to every human infant soon after birth to enable it to learn, especially to learn language.

But how?

The almost innate theory of mind is probably the cumulative product of the iteration of an evolutionary process called the "Baldwin effect," first identified by the psychologist James Mark Baldwin in 1896. In any organism capable of at least some environmental learning, behaviors that are repeatedly rewarded by food or warmth or safety, a bit of prey caught, a mate found, or a predator avoided are reinforced and become more frequent, and behaviors that are repeatedly punished by pain, injury, hunger, or other privation are inhibited and become less frequent; deploying the rewarded behaviors and avoiding the punished ones as a matter of course become learned traits for the organism. Through its life, there will be other traits, most of them hardwired, that can help the organism learn more quickly, deploy the rewarded behaviors for longer and fine-tune them further, and avoid the punished behaviors more often and sooner. All of the traits that help the organism learn will be selected for, over and over again, in the course of evolution. Eventually, selection for traits that accelerate learning will make the organism's ability to learn so early, so fast, and so accurate that it might as well be innate. There's hardly any difference between hardwired species-specific behavior and behavior that is learned early and quickly, needing only slight reinforcement, and that is strongly maintained and operates with great discrimination.

This process of moving originally learned behaviors "inside" to become innate abilities by small steps over countless generations that makes the behaviors easier to learn, earlier learned, further fine-tuned, and more automatically deployed, operates not only for individual behaviors, but also for suites of related behaviors that go together in an organism's life.

All animals are "means-ends" systems. Their behaviors appear to track sources of food, shelter, reproductive opportunities, predators and prey. Since almost all animals are eaten by other animals, it is highly adaptive for all animals to be able to track the means-ends behavior of predators and prey. There will thus be selection for any new capacity that improves on this ability. Tracking the ends-means behaviors of prey or predators requires the ability to track the environmental resources and opportunities those behaviors are responses to—to see what the prey or predators are seeing, smelling, and so on—as well as their ends at any given moment. Then there must be a capacity to combine these two into trajectories that can result in successful predation or escape. In the long-term struggle for survival between predators and prey, anything that facilitates the earliest acquisition of the most accurate version of this ability to predict other animals' ends and means behavior will be selected for. Natural selection driven by Baldwin-effect mechanisms will continue until the suite of related abilities to do this predicting gets as close to being hardwired into cognitive development as biologically possible.

It's the building of such suites of abilities, first in mammals, then in primates, and finally in hominins, that probably produced the might-as-well-be-innate theory of mind in *Homo sapiens*. Here's roughly how this may have happened. Natural selection started with operant conditioning, then used Baldwin-effect processes on learning to produce what the youngest babies and most predatory mammals can do instinctively—predict behavior with a certain reliability: in what direction prey will move next, where edible insects will build their colonies, or birds lay eggs that predators can steal, how predators hunt and what their own pups or cubs might do that endangers them. By the time the great apes appeared, this process had already produced primates who were very good at predicting and acting on a range of behaviors, of their predators, their prey, and their fellow primates.

Primatologists tell us that the great apes have a relatively robust mind-reading ability, robust, that is, compared to most other mammals (Krupenye et al., 2016). Gorillas and chimpanzees are certainly good at predicting the behavior of others. But how they do it is an open question. Do they really carry around in their heads an incipient theory of mind?

That may seem highly unlikely, but there is evidence that at least some chimps (and even some crows) can pass the false-belief test. Researchers had apes watch a video in which a man placed an object in a hiding place, from which it was then removed by an experimenter in a gorilla suit. The man later returned to look for the object. Eye tracking of the kind employed by infants showed that the apes anticipated that the man would look in the place where he'd left the object. This suggested to experimenters that the apes attributed a false belief to the man, an achievement that requires they have some parts of a theory of mind—or at least the concept of a belief (Krupenye et al., 2016).

There is certainly something in ape brains that enables them to deal with other apes, predators and prey, but there are significant limits on their mind-reading abilities (Tomasell, 2014). What's obvious is that they are pretty good at predicting behavior almost from birth. Unsurprisingly, given natural selection's long emphasis on the survival of the fittest, the primates' predictive power seems mainly to be deployed in circumstances of threat and competition. They can't seem to deploy their powers for real teamwork. Even in the case of interaction with close kin, gorillas and chimpanzees only infrequently behave in ways that seem to require much calculation of what the other ape expects and wants. As noted above, the great apes are no good at teaching and learning because they can't imitate. What they lack is the ability that imitation requires to nest beliefs and desires inside each other in their thoughts. Nothing reveals more clearly than this the reasons to doubt that they really do carry around much more than a rudimentary theory of mind. The same almost certainly must have been true for our last common ancestor with them. So what triggered the development of a full-fledged theory of mind in us?

Our full-fledged theory of mind is very likely the result of another Baldwin-effect process, which shapes the great ape's limited, blinkered, rudimentary theory of mind to enhance the payoff from other innate behaviors hominins don't share with the apes. The result was the relatively sudden appearance of the full-blown theory of mind required for hominins to break out of the genetic/population bottleneck that had them barely surviving on the fringes of the savanna for such a painfully long time.

One of these is cooperative child rearing, an innate species-specific trait common to a small number of unrelated species besides *Homo sapiens*: some dogs, elephants, marmosets, tamarin monkeys, and apparently, some hominins, but no other extant primates. Although it's an interesting question why this trait emerged separately and independently in a small number of species, what matters is that it did. As a species-specific inherited trait, it doesn't seem to have been taught and it's hard to suppress. In all of the species in which it emerged, cooperative child rearing has a large effect on social aspects of cognition, for example, eye and gaze tracking, and behaviors that look like teaching/learning, and vocal imitation. And vice versa, the effects feed back on their causes in a coevolutionary cycle (Burkhart, Hrdy, and van Schaik, 2009).

Almost all of these species—dogs, elephants, marmosets, and tamarin monkeys—engage in a suite of behaviors alien to all the great apes, except for us: they "babysit"—they carry, feed, groom, and play with the offspring of others; they share food with them, tolerate crowding by and exchange information with them. Cooperative child rearing would have had increasingly large payoffs once it got started. What it needed first was toleration of others. Put any number of other primates together, and the result is almost always a riot. Apes don't tolerate large groups, or even intimacy beyond grooming. Mother apes only rarely tolerate theft of food by their young. Cooperative child rearing requires tolerance within groups, and, in a coevolutionary cycle, tolerance makes cooperative child rearing easier.

The second thing cooperative child rearing needed was increased ability of caregivers and care receivers to predict and accept one another's behaviors. When a mother wanted to hand one of her young off to another caregiver, the youngster and caregiver had to be ready for the exchange. To make it work well, all three parties to this behavior had to engage in some "mind reading," and all three had to be motivated to participate. Hardwired fitness-enhancing cooperative child rearing would thus have selected for other traits that facilitated it, including ones that made a theory of mind more easily, quickly, and reliably learned by the young.

Gorillas and chimps are very smart. They have roughly a two-year-old's understanding of space, time, causation, and physical object constancy, and they may even have some of the prelinguistic infant's theory of mind. What they lack is "shared intentionality," which requires more of the theory of mind than what they have. When two animals share intentionality, they coordinate their distinct and different individual behaviors to move toward and attain a common goal. In principle, they could do it as the result of operant conditioning over a period of time, but human infants do it without any such conditioning. Experiments show that other primates can't seem to organize themselves, two at a time, into cooperative activities, even when the tasks are simple and the rewards attractive. Nor can they spontaneously help each other or share information with another. Two-year-old children do all of these things—cooperate, share, convey information, and help (Tomasello, 2014). Is shared intentionality by itself evidence enough to attribute a full-blown theory of mind to those who display it? There has to be something in the mind of a human infant that controls the infant's behavior as a function of the behavior of another human with whom it coordinates its movements. It's only because babies who play with others still fail the false-belief test that we don't attribute suites of related beliefs and desires to them—beliefs about the others' behaviors and desires to attain common goals.

Shared intentionality, even if it's only incipient, rudimentary, and limited, when combined with long childhood dependence starts to be selected for. The Baldwin-effect mechanisms that make behaviors more easily learned, and learned earlier, and more automatically deployed give rise to significant increases in fitness. Long childhood dependence provides children both opportunity and incentive to learn. Cooperative rearing of increasing numbers of offspring provides adults with the opportunities to improve and specialize in teaching. Adults most capable of doing so and those of their offspring most capable of imitation and thus learning will themselves have more offspring.

There is considerable synergy in combining even a little shared intentionality with what's already cognitively innate, inherited from our common ancestors and shared with the other great apes: the concepts of causation, object constancy, space, and time. Shared intentionality is indispensable for solving the problem of how to preserve and transmit technologies, such as fire and tool making. Recall that when they can't be easily taught, or when populations are too small, technological innovations too often perish with their makers before they can spread, much less be improved upon. Cooperative child rearing and its partner shared intentionality help solve both problems, with better teaching/learning and bigger populations. They turn the problems of short birth intervals, greater numbers of offspring, and long childhood dependency into opportunities for breaking out of the genetic/population bottleneck hominins found themselves in for a million years or so.

It may have already occurred to you that shared intentionality is almost all that's needed to get language going. That the theory of mind is implicated in language will be obvious to anyone who has noticed the literal meaning of the Spanish and French idioms for "to mean"—*querer decir* and *vouloir dire*—is "to want to say." But "implicated" is far too weak a word. Some linguists argue that language as a human trait emerged, and could only have emerged after a prior and relatively sophisticated if not full-blown theory of mind had already been laid down in our evolution (Malle, 2002).

Start with the fact that mind reading and its precursor—detecting purpose—kick in much earlier in infant development than any overt ability to acquire language. Add the evidence from cognitive science that once it does kick in, infants' acquiring language employs a high level of mind reading: they acquire new words—nouns and verbs—not by associating sounds with the objects or actions speakers are speaking about, but by tracking the speakers' eye movements or facial expressions. Using a theory of mind, infants infer from the speakers' eye movements or facial expressions the referents of the speakers' vocalizations—the words they are using. Saying "rattle" to an infant old enough to learn a new word while you are looking at a ball will get the child to call the ball a "rattle," even if you've just put a rattle in the child's hands and it's dominating the child's visual field (Bloom, 2001; Baldwin, 1991; Baldwin, Markman, and Melartin, 1993).

Now, add to these experimental findings the linguists' assumption that in the Pleistocene, infant language learning had to proceed in the same way, and it clearly follows that no one in the hominin lineage from which we descended learned any words until there was already a fairly sophisticated theory of mind at work.

So the question is, how did a very high level of mind-reading ability get transformed into something like the theory of mind diagrammed in chapter 3 (figure 3.2)? How does that ability get converted into a complicated theory about beliefs and desires and decision making? Like bicycle riding or swimming, mind reading is an ability that can be employed without knowing how it works, much less the theory of how it does, and certainly without being able to put the ability into words. Nor could you teach someone to ride a bike or swim by getting that person to learn a theory. But, the theory of mind is a set of hypotheses and statements employing concepts, some of them quite complicated, like the beliefs and desires that represent, with directions of fit. The theory of mind is something you can express or explain to other people. It's not an ability. The question we need to address is thus how an ability to mind read gives rise to a theory, a set of hypotheses we employ to explain and predict behavior.

Here's how that might have happened: once a certain amount of mind reading more powerful than that of the other primates was in place, proto-language could emerge and then coevolution did the rest: any amount of adaptive communication that exploited the ability to coordinate, in hunting or gathering, imitating and teaching, and dividing labor would select for improvements in mind reading—and vice versa. Once they acquired language and developed it far enough, our hominin ancestors would have the resources to express the results of their mind reading to one another. They would begin to tell stories with plots. And, somehow, along with the stories, the theory of mind would emerge.

It's certainly true that the ability to communicate information and the ability to mind read have two different evolutionary origins and that they certainly could have coevolved. Signal systems between organisms emerged by natural selection alone. Slime mold cells send signals when it's time to clump together and become fruiting bodies. Bees communicate the location of nectar-bearing flowers with their dances in the hive. Lots of mammals have both signal and mind-reading abilities. But if these two abilities were all that was needed for the evolution of a protolanguage and then a theory of mind, we'd expect both to have emerged more than once. Something else was needed to get from mind reading as an ability widespread among even unrelated species to theory of mind as a set of hypotheses employed only by us.

It's tempting to think of that distinctive human capacity, shared intentionality, a level of mind-reading ability beyond anything the other primates manifest, as the catalyst needed to convert mind reading into a theory of mind. Language develops from gestures and vocalizations whose original function—what they were most strongly selected for—was to enable hominins already capable of joint intentionality to fine-tune their behaviors, point out opportunities, agree on tactics, divide the task, get the timing right, and so on. Language was and is the easiest way to get someone else to share your beliefs and your desires, or at least to know what they are, so that they can take appropriate action. We'll come back to this question in chapter 9, once we've reviewed some of the neuroscience of belief-desire pairings.

Two things are important to bear in mind about the process that made the theory of mind as close to innate as it could have become: first, it must have taken at least six million years—that's how far back our lineage began to split off from the other great apes; second, and more important, it involved the coevolution of almost all the processes that operated together to select for the suite of adaptive traits characteristic of *Homo sapiens*: tool making, social and group tolerance and collaboration (to include collaborative child rearing and coordinated hunting), shared intentionality, theory of mind, and the ability to imitate, teach, and learn. Selection for any one of these traits shaped the selection for the others. The whole process was a thoroughly "chicken and egg" phenomenon. The huge problem hominins faced two million years ago and more, was also a huge opportunity. Solving the extinction problem faced by their small population in the long bottleneck was also a chance to move from the bottom of the African food chain to the top.

Once we had a theory of mind, we could begin telling ourselves stories about what we and others were doing, and once we had spoken language, the stories each of us told ourselves could be turned into ones we told one another. And, of course, language gave us words, which we used to improve both the effectiveness of the theory of mind and the detail of its explanations. Eventually, language allowed us humans not just to communicate the results of our theory-of-mind calculations, but also to sharpen, refine, and increase the power of our theory of mind itself. A shared theory of mind made every joint task easier, more efficient, and, most important, far easier to teach and learn.

The progression from a (nearly) innate theory of mind to a fixation on stories—narrative—was made in only a few short steps. We went from explaining how and why we did things in the present, to explaining how and why we did things in the past, to explaining how and why others did things in the present, then in the past, and finally to explaining how others did things with, to, against, and for still others.

Voilà narrative.

As an adaptive trait, the theory of mind is right up there with binocular vision, opposable thumbs, and bipedalism. Together with language, it's what moved us to the top of the food chain. Like other hardwired traits, it's still with us, still shaping our culture, manifesting itself in everything we do. It's still the way we explain ourselves to ourselves and to one another. And because it comes to us before we even learn to walk let alone talk, we aren't just used to it; we can't stop using it, without even thinking about it—automatically, persistently, and unrestrainedly.

But the theory of mind is a theory: a set of hypotheses about human behavior and its unobservable causes. So, as with all theories, questions arise. Is it totally true? Mostly true? Even on the right track toward being true?

Why should we doubt it? Don't we have overwhelming evidence in favor of it? The theory of mind is certainly indispensable to our lives, our institutions, our culture—every culture, for that matter. Why should we doubt it for a minute?

How good is the theory of mind really?

That's a hard question to answer, and an easy one to ignore. Because it's as close to hardwired as a theory can be, the theory of mind is psychologically almost impossible to doubt. It's a theory we are quite unwilling, perhaps even psychologically unable to test. This is one source of the conviction that there is no evidence against it—and a vast amount of evidence for it.

All this uncritical confidence is reason enough to begin at least to consider whether the theory is true, or at least on the right track, mostly true and open to refinement in the direction of complete truth about what makes us tick.

Not everyone has uncritically accepted the theory of mind as even a mostly correct theory of human behavior. In the twentieth century, all the strands of psychological behaviorism rejected the theory either as rank speculation refuted by observation or as "unfalsifiable" dogma impervious to scientific testing. But the behaviorists' attack on the theory of mind failed ignominiously for several reasons. First, their objection to postulating unobservable entities (mental states such as beliefs and desires) was fatally undermined by the example of theories such as those of genetics or particle physics, whose strikingly successful predictions hinged on their appeal to just such entities. Second, the behaviorists could not offer a theory any better than the theory of mind in predicting the human behaviors that most interest us. And, third, many behaviorists appeared to be simply rewriting the theory of mind in their own jargon, or at least to be guided in formulating their hypotheses by the theory of mind they sought to replace. The lesson philosophers drew from the decisive failure of the behaviorists' attack was that the theory of mind was here to stay, indispensable to the explanation of human affairs.

Behaviorism was never very popular. Now we can see why. It flies in the face of a theory that has been bred in our bones. Nevertheless, there was something to the behaviorists' dissatisfaction with the theory of mind.

Indeed, there are compelling reasons to question the theory of mind; I will elaborate on them in the chapters to follow. At the outset, we need to note the theory's Darwinian pedigree is no reason by itself to accept it as true, or even mostly true. The process of natural selection does not as a rule produce true beliefs, just ones that foster survival. Nor does it "track" truths about the world or about us so much as produce "quick and dirty" solutions to problems of survival and reproduction. Quick and dirty solutions to "design problems" gave us the blind spot in each of our retinas and the crossing of our esophagus and larynx that makes us prone to choking on our food. The theory of mind is one of those quick and dirty solutions, a solution wired into our lineage so long ago we couldn't get rid of it if we tried.

Natural selection operates to maximize fitness, but, unless a newly evolved trait has a fitness payoff in the immediate reproductive future of its bearer—say, within a generation or two at most, it will gain no evolutionary traction and won't be passed along in its lineage. Natural selection doesn't wait around for an optimal adaptation either. It selects for the first random variation that deals even slightly better with a design problem. Then, in only a few generations, this trait will become entrenched, at which point it can't be reversed or undone, but only built on or ignored by later variations facing the same or new design problems. But, like natural selection's other quick and dirty solutions to design problems, the theory of mind may have serious defects, and some of these have produced a number of significant imperfections in us.

Among the "imperfections" most relevant here are folk physics and folk biology, which, like the theory of mind, are as near to being innate as they can be. We come into the world with the innate proclivity to adopt, at the slightest provocation, significant falsehoods about nature that, despite their falsity, are convenient to believe, adaptively advantageous, confer substantial short-term benefits, and are extremely hard to relinquish, if at all. Folk biology imposes two mischievous falsehoods on us: first, that we can divide and classify all living things into hierarchical categories with "essential" or defining properties; second, that we should attribute "real" purposes to biological processes generally.

Folk biological taxonomies are almost universal in human cultures across the world. Although they don't seem to be based on generalizing from similarities, they are configured in pretty much the same way regardless of local differences in flora and fauna. We humans appear to formulate such taxonomies even in infancy, hyperactively impose them, and find it extremely difficult not to, although our grasp of them can be disturbed by various brain lesions. Undoubtedly of adaptive benefit, especially in early human evolution, these taxonomies have also been among the greatest barriers to scientific understanding of the biological domain. Since Darwin, biologists have understood that species and other biological categories are not fixed types with defining or essential properties; this insight has been key to advances in genetics and biotechnology. But, even now, we (and this includes biologists away from their laboratories) continue to mistakenly suppose that species, races, and other taxonomic categories are fixed in the way that chemical elements are, that each category has certain essential, defining characteristics, and that certain categories are "higher" and others "lower" on a hierarchy of biological achievement (Atran, 1999). The endemic and persistent racism of all human cultures is testimony to the innateness of folk biology.

More important in folk biology than attributing "essential" characteristics to taxonomic categories is our might-as-well-be-innate proclivity, from infancy onward, to attribute purpose in the biological domain. Well before language, and with the slightest provocation, infants behave in ways that strongly suggest they are employing "teleological" reasoning: things happen in order to attain some end or goal. The experiments suggesting this are the ones employed in early theory-of-mind experiments: eye tracking and dishabituation show the infant to be surprised when inanimate objects appear to track goals or ends, and when animate objects fail to do so (Gredeback and Melinder, 2011). We noted above that there will be strong selection for treating prey and predators as ends and means systems. The ability to do this is so critical to survival that Baldwin-effect processes will make it as close to innate as biologically possible. In humans, the ability to track ends and means patterns expresses itself in the indiscriminate attribution of purposes to almost all natural processes or their causes, a proclivity shaped and sharpened into a theory of mind by the age of three or four.

Even more than the "essentialist" taxonomies of folk biology, the mistaken belief that there are purposes in the biological domain has obstructed scientific progress since long before Aristotle. This belief was not fully recognized for its falsity even when Darwin showed there are no real purposes in nature, replacing them with a purely causal mechanism—random or blind variation and natural selection. Newton excluded purpose from the domain of physics, but it remained fully in force in the biological domain until 1859, when *On the Origin of Species* first appeared.

The connection between teleological thinking in biology and the theory of mind had been evident for thousands of years. The argument from design for God's existence is an inductive inference from the appearance of purpose in the biological realm to the best explanation for its emergence: the theory of mind applied to God. It was the benevolent desires and the infallible beliefs of the deity that worked together to realize all the harmony of means to ends we detect in the biological domain (Paley, 1802).

Once Darwin identified a purely causal mechanism that produces the appearance without the reality of design, the role for teleology and the scope for the theory of mind in biology evaporated. After 1859, the explanatory and predictive hold of the teleology-driven theory of mind remained only in the domain of human affairs. Nevertheless, the thoroughly false teleological conception of the biological remains a fixed proclivity of the human mind, from infancy it would seem.

Natural selection has laid down and shaped several other highly adaptive falsehoods in our human minds. These false beliefs seem to have worked out well on the African savanna for a million years or so, and even today they don't get us into much trouble in our everyday lives. But two entire disciplines at the intersection of economics and psychology have arisen to identify them: behavioral economics and decision science. These disciplines have uncovered the rules of thumb we employ in our everyday decision making, rules that violate mathematically and empirically well founded principles of rational decision making (Gigerenzer, 2011). Nobel Prizes have been won identifying and explaining these cognitive biases (Kahneman, 2011). For example, people often commit the gambler's fallacy of assuming that a run of coin flips producing four heads increases the probability that the next coin flip will produce tails. They greatly overestimate the risks of air over ground transportation, largely in response to the publicity accorded aviation disasters, however rare they may be.

The explanations cognitive scientists typically offer for the rules of thumb that produce these false beliefs invoke their evolutionary pedigree. Faced with decision problems, natural selection selects for quick and dirty (fast and frugal) rule-of-thumb solutions, ones that have a good chance of getting the decision right in most cases and are easy to either teach or hardwire, ones that don't require a great deal of language or highly quantitative concepts to employ (Gigerenzer, 2011). As humans pass into the Holocene (about 10,000 years ago), then the advent of agriculture, feudalism, the Industrial Revolution, the mistakes these rule-of-thumb solutions produced become more and more costly. Only now have these rules-of-thumb at last been discarded for more reliable ones in diverse areas of science and business. They nevertheless remain with us in our everyday lives, where they continue to be relatively harmless.

In its response to particular environments, natural selection often appears to overshoot and then sometimes to fine-tune. But what looks like overshooting may really be adaptation to a different "design problem," or even to a new problem created by the solution to an earlier one. Our theory of mind is probably embroiled in one such tangle of "problems" and "solutions." To encourage our use of the theory of mind, natural selection has selected for those of us who are rewarded with a feeling of satisfaction, an "aha" experience, when we do (Gopnik, 2000). Neuroscientists have shown that hearing a story, especially a tension-filled one in which the protagonists' emotions are invoked, is followed by the release of pleasure-producing hormones such as oxytocin, which is also released during orgasm and is important for sustaining cooperation as well (Zak, 2015). The pleasurable sensation reinforces our subsequent use of the theory of mind and is another source of our love of history and biography.

But when natural selection endowed us with an adaptive theory of mind and rewarded us with pleasure for using the theory, it seems to have overshot, turning us into "hyperactive agency detectors," who suspect conspiracies and motives behind everything that happens at all. You can't really fault evolution, though. It was hard enough and took long enough for it to shape a working theory of mind—six million years if you start from our last common ancestor with the other living primates who lack one. Imagine how much harder it was to shape a full-blown theory of mind and to include in it a hardwired restriction on its application: "Apply to humans and to their predators and prey only, please." In fact, making us hyperactive agency detectors might not have been overshooting at all. It might instead have been something akin to shrewd cost-benefit analysis. Better for our survival if we made what scientists call "type II errors"—suspecting threats or dangers when they weren't there. After all, in the Pleistocene (some two million years ago), the costs of being wrong about threats or dangers were low, but the benefits to being right were high.

We humans have been attributing intentions and motives to purely physical processes and phenomena since time immemorial. Think about our proclivity to see hurricanes, cyclones, earthquakes, tsunamis, floods, and other natural disasters as the purposeful products of an all-powerful agent. Indeed, religion may well have been, at least to a significant degree, the result of overshooting in the use of a theory of mind. But here natural selection may have taken advantage of an overshoot to solve another design problem, one that has been obvious to evolutionary anthropologists since Darwin made it explicit in *The Descent of Man* in 1871.

Once our ancestors moved beyond working with just their kin, the problem of cooperation became serious since the short-term benefits of free riding, cutting corners, and cheating were so great. Remember, natural selection selects only for the short term. When out of sight of others, the temptation to break rules would often have been overwhelming. What was needed was an enforcement device that worked under these conditions. Well, cognitive social psychologists have found good experimental evidence to show that the belief in an all-powerful but invisible or otherwise unobservable "watcher" or "punisher" would most likely have kept people on the "strait and narrow" (Bateson, Nettle, and Roberts, 2006). Accordingly, Darwinian cultural evolution, in which practices were selected for according to their adaptive value, would likely have established religions incorporating a morality-enforcing deity. One good way of doing so is by exploiting the design flaw of hyperactive agent detection (Boyer, 2008).

Now we see why we love stories—plot-driven narratives—why we crave them, are addicted to them, need them. Our minds are hardwired to impose the theory of mind on chronologies—to make them into narrative histories—and to find pleasure in doing so (Gopnik, 2000). We humans are very much the products of a natural selection process that has made us seek out such histories and conduct our affairs by employing them.

The trouble is, the theory of mind is not a very good theory, not at least by the standards we set for theories everywhere else. This should come as no surprise. If the theory was, in effect, conferred by natural selection, and not the result of careful scientific investigation, it was not likely to be anything more than a relatively recent quick and dirty solution to an evolutionary "design problem" or perhaps a quick and dirty response to a "competitive opportunity," like most of our other inherited traits.

The theory wasn't really "designed" as anything more than a tool for anticipating the immediate future behavior of friends or foes, cooperators or competitors, predators or prey, mates and offspring, kith and kin. And it wasn't really very good at predicting immediate future behavior, unless you shared a lot of the same environment and experience with the people whose behavior you were trying to predict. Even then, you couldn't predict the details of their behavior much further out than the next few minutes or hours in any detail. Of course, in most cases, detailed long-term prediction wasn't really necessary back in the Pleistocene.

But the theory of mind was the first, indeed the only, cognitive tool available for coping with other people when human life became more complex, and it was already wired in, hard to give up, and hard not to think things through with. Which meant that all of subsequent theorizing about human affairs has been built on its rickety foundations. And not just theorizing but all the different institutions, practices, norms, conventions, codes, laws, etiquettes, manners, and styles that characterize our diverse cultures as *Homo sapiens* have been built (sometimes by explicit design, usually not) employing the theory of mind. And sometimes these "constructions" regulated our behavior well enough to make the theory of mind an increasingly reliable tool of longer-range prediction. Once the institutions that structure human affairs got up and running with the advent of agriculture, the theory of mind became especially good at predicting what we wouldn't do, or at least usually not do: violate the rules that constituted these social constructions. By giving people incentives to abide by and, even more important, disincentives not to break these rules, cultures, societies, and polities were able to make reliable predictions of what, by and large, we wouldn't do. But, as to what people *will* do, the theory hasn't gotten much better at prediction than it was in the Pleistocene. More than one thinker has recognized how pervasively the theory of mind underlies the culture we've built from our Pleistocene beginnings. The well-known philosopher of psychology Jerry Fodor once expressed the fear that if the theory of mind turned out to be mistaken, the result would be "the greatest intellectual catastrophe in the history of our species" (Fodor, 1987, p. xii). Actually, he needn't have worried. We can't give up the theory of mind for most practical purposes, even if it turns out to be mistaken. And since we can't, the precious institutions built on its rickety foundations are in no more danger of collapsing than the theory itself is of being given up.

But what makes the theory so rickety?

For one thing, it's a lot different from all the other theories we've ever come up with. It's different in several respects, few of them reassuring about its truth, even its partial truth. It doesn't look like it's anywhere near any theory that can make improvements on its explanatory and predictive powers. And this despite its being the oldest theory we've got, and thus one we've had countless opportunities to improve on. But we haven't improved on it, not in all that time.

Think how long we've been using the theory of mind explicitly to explain and predict. Homer employed it in *The Iliad* (ix:488–505) to explain why Odysseus left Troy to return to Ithaca, and we understand his explanation perfectly well almost 3,000 years later:

I say no wealth is worth my life! Not all they claim
was stored in the depths of Troy, that city built on riches,
in the old days of peace before the sons of Achaea came—
not all the gold held fast in the Archer's rocky vaults,
in Phoebus Apollo's house on Pytho's sheer cliffs!
Cattle and fat sheep can all be had for the raiding,
tripods all for the trading, and tawny-headed stallions.
But a man's life breath cannot come back again—
no raiders in force, no trading brings it back,
once it slips through a man's clenched teeth.
Mother tells me,
the immortal goddess Thetis with her glistening feet,
that two fates bear me on to the day of death.
If I hold out here and I lay siege to Troy,
my journey home is gone, but my glory never dies.
If I voyage back to the fatherland I love,
my pride, my glory dies ...
true, but the life that's left me will be long,
the stroke of death will not come on me quickly. (Homer, 1998, p. 265)

A thousand years or so before the *Iliad*, the Epic of Gilgamesh employed the same theory of mind.

We still read the fourth-century B.C. Greek historian Thucydides's explanation in *The History of the Peloponnesian War* (iv.14.2–3) because his theory of mind is almost[2] exactly the same as ours:

The Spartans, distressed at the disaster because their men on the island were being cut off, rushed to the rescue, and going into the sea with their heavy armour, laid hold of the ships and tried to drag them back; everyone felt that the action was

2. "Almost." About the only difference worth noting is the introduction of psycho-babble from the beginning of the twentieth century onward, whether by Freudians, behaviorists, cognitive social psychologists, or behavioral economists. Many concepts have been added to the lexicon of the theory of mind, almost all of them synonyms for "beliefs" and "desires," "habits" and "emotions." These additions have fostered the illusion of increased understanding while doing nothing much to further the theory's predictive power. As evidence of this, consider how little better at prediction the marketing departments of businesses or even of Ivy League business schools have gotten in the last century. With millions on the line, it's all still a matter of guesswork.

hindered if he was not himself involved in it; great was the disturbance, and quite unlike the naval tactics usual to the two sides. For the Spartans in their eagerness and consternation were so to speak fighting nothing other than a sea battle from land, while the Athenians victorious and wanting to push ahead as far as possible in the good fortune of the moment were fighting a land battle from ships (qtd. in Rood, 1998, p. 7).

The theory of mind has stood the historian's test of time, our own time and the *longue durée* of human history. We have used it to explain ourselves to others, to predict some of the behavior of other people in whose company we find ourselves, including total strangers, even people from foreign cultures.

But there is something odd and disturbing about the fact that it's almost exactly the same theory now that it was in Homer's time. Unlike every other theory that's been around for a long time, the theory of mind has not gotten any better. That it's still pretty much the same in content, explanatory range, and predictive precision as it was four thousand years ago couldn't be because it was already finished, complete, and in need of no improvement in the Bronze Age.

Even a theory that was originally bred in our bones should be one we can improve, sharpen, and make more predictively precise as a result of experience, observation, and experiment. This, after all, has been the fate of every other theory of significance that we've inherited or acquired. We've been able, to refine even primitive theories like folk physics and folk biology and eventually to replace them with far more powerful, accurate, useful theories. But not the theory of mind.

A scientific theory is set of hypotheses, guesses that get tested, revised and improved over time, or that, if falsified or refuted, get overturned, giving way to those of a new and improved theory. Every other theory, even the most successful, eventually gets improved or superseded. Nothing like this has happened to the theory of mind in all the millennia since it first emerged in human culture.

If the theory of mind is an explanatory and predictive theory, then there ought to be at least some circumstances under which we would have revised, improved, or even given up on it. Almost no such circumstances seem to have arisen to the present day.

The qualification "almost" is grudgingly added to accommodate a brief nineteenth- and twentieth-century attempt to treat the theory of mind as

Plate 1
Punch and Judy show. From https://en.wikipedia.org/wiki/Punch_and_Judy#/media
/File:Swanage_Punch_%26_Judy.jpg.

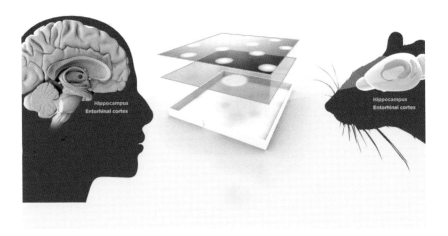

Plate 2
Relevant parts of the rat and the human brain—the hippocampus and the entorhinal
cortex. From https://www.extremetech.com/wp-content/uploads/2014/10/nobel-prize
-grid-cells-diagram-place-cells-rat-human-640x353.jpg. With permission from Mattias
Karlén/The Nobel Committee for Physiology or Medicine.

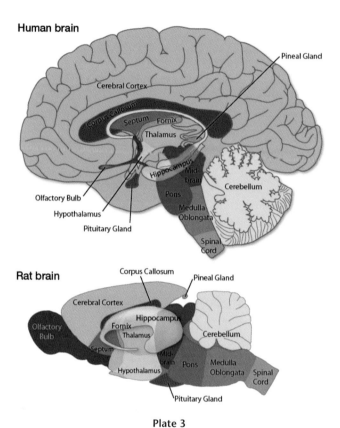

Plate 3

Homology of the rat and the human brain. From http://learn
.genetics.utah.edu/content/addiction/mice/brains.jpg.

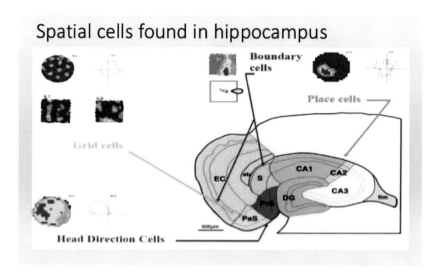

Plate 4

Different kinds of dedicated cells in the hippocampus of the rat brain. Grid cells, head direction cells, and boundary cells all feed into or have "projections" to the place cells; all these cell types in the rat brain hippocampus are shared by the human brain hippocampus. From O'Keefe, 2014, fig. 2.

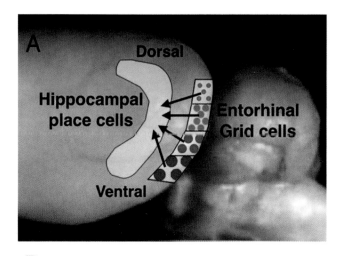

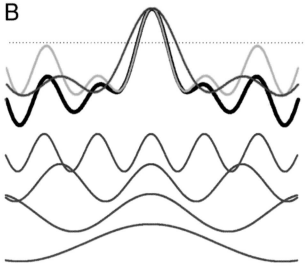

Plate 5

Different-sized dots identify the grid cells sensitive to larger and larger spaces in the rat's environment. The waves represent "theta wave" oscillations from the grid cells that combine at the place cells to locate the rat. From Kubie and Fox, 2015, fig. 2.

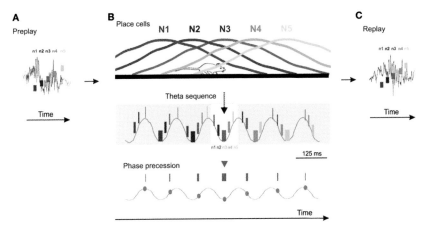

Plate 6

Cartoon model of theta sequences and phase precession in the CA1 area of the hippocampus. Preplay and replay panels A and C will be important later in chapter. The theta sequence and phase precession diagrams illustrate how place cells carry information about the rat's location—by firing rate of distinct cells and by when they fire in a theta oscillation. From Dragoi, 2013, fig. 1.

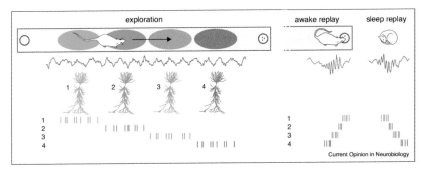

Plate 7

Theta waves are drawn below the track, and successive place cell circuits in the hippocampus are excited as the rat moves through different places in the track, with signature firing patters. These are recapitulated in away reverse replay starting with reward location and then recapitulated in forward replay during sleep for consolidation. From Girardeau and Zugaro, 2011, fig. 2.

Plate 8

Deep blue, the IBM chess champion. From https://i.amz.mshcdn
.com/GH65hG_jpZtWm9J_iNF9QCD0Rak=/http%3A%2F%
2Fa.amz.mshcdn.com%2Fwp-content%2Fuploads%2F2016%
2F02%2Fkasparovdeepblue-19.jpg. Courtesy of Getty.

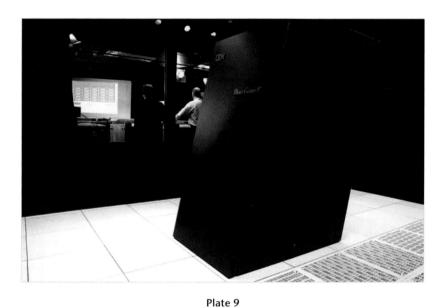

Plate 9
Part of Watson at the IBM Jeopardy! studio, https://static01.nyt.com/images/2010
/06/20/magazine/20Computer-span/20Computer-span-articleLarge-v2.jpg. Courtesy
of New York Times.

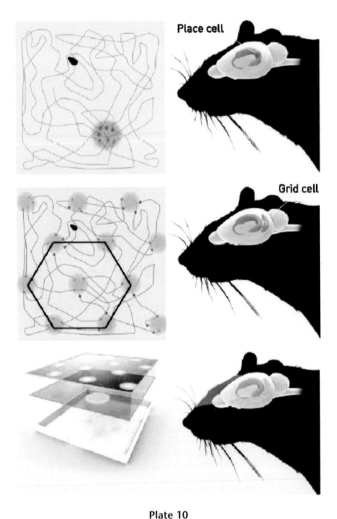

Plate 10

Place cells in the hippocampus and grid cells in the medial entorhinal cortex of the rat's brain. From http://www.frontline.in
/multimedia/dynamic/02171/fl14_nobel_medicin_2171449g
.jpg. Courtesy of Mattias Karlén/The Nobel Committee for
Physiology or Medicine.

a precursor to a scientific theory that was to be a more precise, formally stated, mathematically tractable set of hypotheses about quantitatively measurable versions of the theory of mind's crucial explanatory variables. The new theory was advanced by the "marginalist" economic theorists of the late nineteenth century and their twentieth-century successors, microeconomic theorists, who substituted "preferences" (originally measured in "utiles") for "desires" and "expectations" (eventually measured in quantitative probabilities) for "beliefs." The new theory combined them, in the same way the old theory of mind dictated, to make decisions that determine choices. One influential economist of the late nineteenth century, Francis Ysidro Edgeworth, developed this formalization of the theory of mind in a 150-page essay, *Mathematical Psychics*.

But when economists came to realize that, formulated with precision, the new theory wouldn't work to predict the actual economic behavior of individuals beyond what the commonsense theory of mind had been able to do, they gave it up as an explanatory theory, too, denying any interest in the psychology of economic choice. Instead, most twentieth-century mainstream economists defended rational choice theory as a convenient instrument for organizing aggregate data about entire markets made up of large numbers of buyers and sellers. Some influential economists even recognized that the theory works that way because competitive markets impose a sort of Darwinian "survival of the fittest" constraint on buyers and sellers that selects for optimizing behavior, regardless of what psychological processes produce it. This insight was eventually exploited by game theory and experimental economics in ways that had important influences on history too, as we'll see in chapter 12.

Economics has quite a bit of cachet for being more mathematical and therefore more "scientific" than the other social sciences. Political scientists, sociologists, anthropologists, and social psychologists alike were impressed with the clarity of economics' formalization of the theory of mind as rational choice theory, but, unlike mainstream "neoclassical" economists, at least some of them were interested in explaining the behavior of individuals, not just large groups or aggregates of them. When, however, they put their use of rational choice theory to the test, the results were "mixed," to say the least (Kahneman and Thaler, 2006). For example, when specialists in international relations began with the assumption that governments or their leaders were rational, they predicted that wars would never happen (Morrow, 2011).

The predictive failures of a rational choice theory of mind led to creation of a new discipline, behavioral economics, in the late twentieth century. Cognitive social psychologists, and especially a new generation of behavioral economists, sought to explain these predictive failures through a variety of devices. Sometimes they advanced hypotheses about systematic human mathematical mistakes or imperfect shortcuts in reasoning that substitute for intricate calculations (Gigerenzer, 2011). Sometimes they suggested that people's desires, and especially their aversion to losses, were not as straightforward as economists had assumed (Kahneman and Thaler 2006). Discoveries about "bounded" (i.e., limited) rationality have explained a lot about the limitations of rational choice theory. Alas, as accounts of individual decision making, the behavioral economic models motivated by this research program have proved little better at predicting individual choice than Homer and Thucydides were.[3] The chapters to follow will show why behavioral economics hasn't been able to improve the theory of mind, despite its commitment to the scientific method.

All this suggests that the theory of mind is not much of a theory at all. Indeed, it seems more like a crude instrument than a proper theory, more like a tool just good enough to get us up the food chain on the African savanna. But if that's all it is, why hasn't it been superseded by better theories the way folk physics and folk biology have?

Is the glass half full or half empty? Even though the theory of mind is no better at prediction than it was at the dawn of written history, it's pretty good at predicting the behavior of people in our vicinity in the immediate future. Surely it can't be that good by accident? And if its limited predictive success is no accident, its explanations must be mostly or at least sometimes right, mustn't they?

The trouble with this argument becomes obvious once we reconsider folk physics and folk biology. Both were predictively reliable enough to get

---

3. Nor have neoclassical economists been able to systematically incorporate the insights of behavioral economics to improve their predictions about market or aggregate behavior either.

us all the way to the scientific revolution of the seventeenth century, even though their explanations turned out to be not even mostly correct. It's worth recalling here that "folk psychology" is the label most philosophers employ for the theory of mind.

Why hasn't folk psychology—the theory of mind—suffered the fate of folk biology and folk physics? It's easy to say that, unlike them, we just haven't been able to find any alternative to the theory of mind. But that simply raises the question, why not? Evolutionary anthropologists have explained why we can't give up the theory of mind as a psychological disposition: it was selected for so long ago. But that doesn't mean we can't find an alternative to it as a scientific theory.

If we're going to continue to endorse the theory of mind as providing real explanatory knowledge, we're going to have to present arguments for its being mostly true, for its being on the right track, for its claims to successfully identify the causes of our actions in the beliefs and desires it invokes to explain them.

Before we begin trying to show the theory of mind is right, it's worth reviewing why our commitment to narrative history requires that it's at least approximately true. Chapter 2 presented narrative historians' contention that the kind of knowledge their histories provide need not pass even the weakest demands for predictive success. Narrative history's job is to tell us the facts about what happened in the past and why it happened, with no eye to any payoff for the future. Chapter 3 made clear the role a theory of mind plays in the explanations narrative history offers, and chapter 4 and this chapter shed light on how we acquired the theory of mind. Vindicating narrative history's explanations requires that the theory of mind be at least mostly true, on the right track to the correct account of the actions and events human history (what happened in the past) consists in. To be even an approximation to the truth, the theory of mind must do two things: first, it must identify the actual causes of the events in human history; and then must get right the connections between these causes and their effects in the actions and events that make up human history. Can it do these two these two things?

Both common sense and our conscious awareness of the role of beliefs and desires in our actions seem to cry out that the answer to this question can only be "Yes!" In fact, introspection alone makes us feel it's preposterous, absurd, not even worth discussing whether the theory of mind could be wrong about whether beliefs and desires are the causes of our own decisions and actions.

But in chapter 6 we'll see why history demands that there be a fact of the matter about whether beliefs and desires bring about what its narratives explain. Then we'll see why our conscious awareness that beliefs and desires work the way narrative history requires they do is sheer bluff.

6 What Exactly Was the Kaiser Thinking?

Historians report that in 1914, with most of the world already plunged in war, Prince Bülow, the former German chancellor, said to the then Chancellor Bethmann-Hollweg: "How did it all happen?" And Bethmann-Hollweg replied: "Ah, if only one knew."

—John F. Kennedy, October 19, 1963

Prince Bülow's question was not about the sequence of events that began with the assassination of the Austrian archduke in Sarajevo in late June 1914 and culminated in the German violation of Belgian neutrality on August 4. It was about which decisions, choices, and actions made the war happen in the first place. Most historians agree that it was the "blank check" Kaiser Wilhelm gave to Austria-Hungary that precipitated the First World War. But the historical narratives disagree about why he decided to go to war. And historians continue to argue about why he did to this very day.

Some favor a simple explanation: they cite documents that show that Kaiser Wilhelm (figure 6.1) had decided on a war of aggression as early as December of 1912, to begin as soon as the German navy was ready.

According to these narratives, by July of 1914 Wilhelm believed that his navy was ready, at which time he gave Austria-Hungary a "blank check" to start the war. Most historians find this explanation far too simpleminded. Noting that Wilhelm had given up his battleship arms race with the British in 1913 and shifted military expenditures back to the army, some historians argue that attributing such a small and simple set of beliefs and desires to the German kaiser ignores his long-standing desire to maintain the influence of Austria-Hungary as a European power and his belief that to do so he had to back its leadership in the Balkans. Other historians explain that the kaiser provoked the war in 1914 owing to his desire to prevent Russia from

Figure 6.1
Kaiser Wilhelm as admiral in 1912. Was he just waiting till his fleet
was ready? https://upload.wikimedia.org/wikipedia/commons/6
/63/Bain_News_Service_-_The_Library_of_Congress_-_Kaiser
Wilhelm%28LOC%29_%28pd%29.jpg

becoming the dominant power in the Balkans and his belief that delay in starting the war would result in a stronger, modernized Russian army. Other historians explain that the kaiser acted on his desire to prevent the growth of democratic forces in Germany and on his belief that a war to expand Germany's territory would unify the country politically. Still others explain that what Wilhelm really wanted was to break up the Triple Alliance of Great Britain, France, and Russia, and thought giving Austria-Hungary a "blank check" to go to war against Serbia would bring Russia into the war without Britain and France, thus bringing an end to their alliance.

Each of these four explanations is easy to understand, none requires special further information, and all make sense of Kaiser Wilhelm's actions. Since all four are compatible with one another, they could all be correct, all wrong, or some combination of them could be right. In writing their narratives of what brought about the war, historians have to choose. But a hundred years of "historiography" since 1914 has failed to settle the question of which explanation or combination of explanations is right, which explanation should be accepted by those who seek to know why the war began. Is there a way to decide?

All of these explanations are consistent with the extant evidence. There might at one time have been clear evidence for or against some or even all of them, but, if so, it no longer exists. As noted, the explanations could, of course, all be correct to some extent or other. Does it matter? Do we need to decide? Well, if history is simply about telling a good story, we probably don't. Any of the four or all of the four would satisfy our curiosity the way a good narrative explanation is supposed to.

Historians are still arguing over the matter because history is supposed to be more than simply a good story. It's supposed to give us the best and best-informed guess about the "fact of the matter"—in this case, about what the kaiser was actually thinking that drove him to provoke the war.

We may never know which one or combination of the four belief-desire pairings—if any—led to the kaiser's actions. Maybe another quite different belief-desire pairing actually motived him, and no one has ever figured it out. Still, there has to be a fact of the matter about what the kaiser actually believed and desired, even if we can't ever establish it. Without such a fact of the matter, no narrative of why the kaiser gave Austria-Hungary a "blank check" can be correct. All the narratives purporting to explain how the war began will be wrong. Without facts of the matter for what people believed and desired, narrative history will turn out to be not knowledge of why things happened in the past, but mere storytelling, no different from historical fictions all the way back to Homer's epics.

But hold on. Surely, we have excellent reason to suppose that there is a matter of fact here about which pairing of beliefs and desires led to the kaiser's actions, even if we'll never know enough to identify that pairing conclusively? We know that there must have been such a pairing just by thinking about our own actions, and the pairings of beliefs and desires in our own minds that lead us to take those actions. Surely, most of the time we understand the real causes of our own actions. We may sometimes get the facts wrong, misremember, or delude ourselves later about why we did something, but, at least sometimes, in fact, most of the time, we get it right, especially at the moments we're deciding, choosing, or acting. At those moments, we know that there is at least one narrative that must be right (perhaps incomplete but at least mostly right). We know it from our own introspection about our own actions. We do things because of beliefs and desires; we know how the process described by the theory of mind works in our own minds, which gives us confidence that it works to explain other people's behavior. The process described by the theory of mind works "transparently" in our own heads, doesn't it? Our confidence in historical narratives as, at least in principle, capable of providing explanatory knowledge is just a matter of generalizing from our own introspective certainty. But what exactly does introspection tell us we know about our own beliefs and desires?

We know that we have beliefs and we have desires, and that they cause us to act in the way the theory of mind tells us they do. Each of us can see the beliefs and desires with our mind's eye and can see the causal link between them and action in the same way, by conscious introspection.

Or can we?

Can introspective conscious awareness by itself vindicate these assumptions—that we have distinct, definite or indistinct, indefinite) beliefs and desires, and that they cause actions in the way the theory of mind tells us? We suppose that it can, we think so fervently. But few, if any, of us ever "enter into" (David Hume's words)[1] ourselves carefully enough to catch a

---

1. As Hume writes in *A Treatise of Human Nature*: "For my part, when I enter most intimately into what I call *myself*, I always stumble on some particular perception or other.... I never can catch myself at any time without a perception, and never can observe anything but the perception" (Hume, 1985 [1739–1740], p. 180; emphasis original).

glimpse of what our beliefs and desires consist of, and how they work to control our movements and actions. When we do, it becomes obvious that introspection can't ground the theory of mind at all.

Start with the obvious claim that what we desire and believe brings about what our bodies do. How do we know?

Here's a classic experiment the famous seventeenth-century British philosopher John Locke asked us to perform. Decide to raise your arm. What happens? You have the experience of wanting or willing your arm to rise, followed by the arm going up. Why did the arm go up? Because you wanted it to. Obviously, the act of will that you were conscious of brought about the movement of the arm. And you were conscious that it was the act of will that caused the arm movement, right? Locke used this argument to show we are aware in consciousness of the causal link between what happens in our thinking and what our bodies do, just as the theory of mind requires.

Fifty years after Locke wrote, however, David Hume showed what was wrong with this argument and with introspection's claim to knowledge about the causes of human action (Hume, 1985 [1739–1740]). When you think about the experience carefully, all you ever notice is the feeling of deciding to raise your arm, and then slightly later, your arm going up—a bit of introspective awareness followed by a "kinesthetic sensation"—the feeling that accompanies a bodily movement. That such a sequence is not reliable as a source of knowledge we now know from the data about phantom limbs. An amputee with phantom limb syndrome has the experience of willing an amputated arm or leg up, followed by the sensation of the limb actually rising, even though it's not there to rise. Hume didn't need that observation to notice that, in the normal case of what we think is one thing causing another, all we ever have is one thing followed by another, and, in the case of introspective awareness, all we ever get is correlation.

When a conscious desire coincides with the action desired—the desire to raise a limb with the limb rising in the case above—it could be causation, or it could be that both desire and action are effects of some prior quite unconscious event or process. Whichever it is, the subjective experience doesn't uncover any mechanism that links the felt desire to the kinesthetic sensation of limb movement, still less to the physiology of muscle contraction that the limb's movement entails. When the theory of mind tells us it's desires that bring about movements of the body, it's making a claim that might be right but that, as Hume showed, conscious experience alone doesn't provide sufficient evidence for.

If it could be shown that what happens in the part of the brain where the desire to raise a limb is experienced actually sets off a neural process that culminates in the flexing of muscles that make the limb rise, then the theory of mind's causal claim would be vindicated. It took another 230 years after Locke for neuroscientists to do the experiment that would show us whether this was in fact so. Alas, things didn't work out the way Locke supposed. It was in the 1980s that Benjamin Libet's experiment (Libet, 1985) showed that the conscious decision to do something as simple as raising a limb actually occurs *after* the unconscious events in the brain send the signal to the limb to rise and *before* the signal gets there—a finding consistently replicated by neuroscientific experimenters over the next thirty years

The conscious feeling of wanting, willing, deciding to raise a limb is just a side effect, a by-product of the unconscious neural event that is enough to get the limb in motion. But because common sense and conscious awareness are so prone to the fallacy of *post hoc, ergo propter hoc* even after learning about the Libet experiment, it's impossible for us to shake the conviction consciousness provides that there is a direct causal link from desire to action of the sort the theory of mind demands.

Some have inferred from the results of Libet's experiment and its replications that there is no free will since the conscious experience of willing can't be the cause of the limb movement. This inference is almost certainly overhasty. After all, free will doesn't require consciousness. For the same reason, it would be equally incautious to infer that the theory of mind is refuted by the Libet experiment. The theory of mind doesn't require that all, most, many, or even any beliefs and desires be conscious, so it's compatible with causal irrelevance of consciously experienced desires. What can't be denied is that the conscious experience of willing a limb to rise is not a link in the chain of events that brings about the limb's going up. It's just a side effect.

Devastating as it is to the claim that introspection is generally reliable, this result drives a stake into the heart of the claim that the theory of mind is grounded in consciousness. However right the theory of mind may be about beliefs and desires as the causes of human action, there's no special evidence for it in our conscious experience. Nor does introspection guarantee the link between desires and beliefs, on the one hand, and what we do, on the other.

To review: we started out with the question of why the kaiser went to war. We saw that we might never be able to answer that question, but that at least we could be confident there were reasons why he went to war, there was some matter of fact about his beliefs and desires that explained the "blank check" he gave to Austria-Hungary. Our confidence about that matter of fact was grounded in our own conscious awareness that we act out of desires and beliefs in the way the theory of mind describes.

Now we've seen that conscious introspection doesn't provide grounds for confidence about the causal link the theory of mind describes. Well, as a source of evidence for the theory of mind, consciousness faces an even more serious problem, one that is hard to take seriously the first time you face it—and hard even to understand what it might be. But, once you see the problem, you'll begin to worry whether any narrative, historical or otherwise, can be trusted. The problem starts out as the question of how we know what our thoughts mean, what our beliefs and desires are *about*. That turns into the problem of whether thoughts are anything like what consciousness seems to tell us about them. And the problem then becomes whether there is really anything in our minds that measures up to what needs to be there for the theory of mind to do any explaining at all. What exactly are the beliefs and desires that we're consciously aware of?

Recall the diagram used in chapter 3 to sketch out the theory of mind (figure 3.2). The statements in the belief and desire boxes are what makes them beliefs *about* how some aspects of the world are arranged and desires *about* how some aspects of the world might be rearranged. Beliefs *represent* the way things are and desires *represent* the way we want them to be. Both have directions of fit: beliefs are supposed to fit the world; desires are supposed to have the world fitted to them. In your conscious awareness, you can "read" the content of a belief: it's usually something like a sentence in silent speech, like the belief that "Paris is the capital of France." It's a little harder to read the content of a conscious desire as a sentence. You need to "rewrite" the desire slightly, from "the desire to be in Paris tomorrow" to "the desire that I am in Paris tomorrow." Beliefs and desires are *propositional attitudes*. This is how beliefs and desires get their direction of fit: beliefs are attitudes about the truth or falsity of propositions, attitudes about how we believe the world is or isn't arranged, say, the belief that Napoléon conquered Russia. Desires are attitudes about how we want the world to be arranged, say, the desire that Napoléon conquer Russia. We express the *content* of our beliefs and desires in sentences—silently to ourselves, out loud to others. The sentences describe the things our beliefs and desires are *about*—Paris or

Napoléon, Russia or conquest. The beliefs and desires *represent* the way we believe the world is and the ways we'd like it to be. These three things—*representing, having content,* and *being about things*—are crucial to what beliefs and desires are. They are the three most distinctive things about beliefs and desires because they give them their explanatory roles in the theory of mind: the outcomes that our desires are *about* together with the means to attain these outcomes that our beliefs *contain,* make sense of, explain our actions.

I am belaboring the obvious because these are the features of beliefs and desires that we need to understand if we are to vindicate their role in the theory of mind. The theory works by focusing on the *content* of our beliefs and desires, what they are *about,* what they *represent.* In short, what they *mean.*

Of course, it's our immediate conscious awareness that guarantees that beliefs and desires have these three features. We can see them with our mind's eye or hear them with our mind's ear every time we bring a belief or a desire to consciousness. But what exactly are we seeing with our mind's eye or hearing with our mind's ear? What does the "aboutness," content, meaning, representational character of a belief or a desire consist in, what is it about a desire or belief that our introspection makes us immediately aware of? To answer this question, let's go back to chapter 3 and consider again how street or road signs have content, represent, and are about when to stop, go, yield, merge, and so on.

Think about the red octagon and inverted yellow triangle street signs you see at intersections. Their shapes and colors alone express the one-word sentences "Stop." and "Yield." The red octagons are *about* stopping; they represent the imperative one-word sentence commanding you to stop. The yellow triangles do the same for yielding. But there is nothing intrinsically stop-commanding about the red octagons or yield-commanding about the yellow inverted triangles. What each street sign means depends entirely on how we *interpret* it, what *content* we give to it. The signs get their meaning "by convention." If back in Detroit in 1915 when and where street signs were decided on, the inverted yellow triangle had been chosen to mean "Stop" and not "Yield," it would have meant what the red octagon now means.

It's convention that makes red octagons into *symbols* that mean "Stop" and yellow triangles into symbols that mean "Yield." What makes them symbols is that people interpret them to mean the commands that are expressed in English by those two words.

Now the question we need to answer is what gives the content or meaning to the thought in our consciousness that we should stop or yield here? What makes any thought in our consciousness *represent* what the thought is *about*?[2] One thing we need to get completely clear on from the outset: the images and silent sounds that are the thoughts in our consciousness do *not* have their *content* the same way that symbols like red octagons and yellow triangles have *content*. Interpretations, conventions, and decisions confer meaning, *about-ness*, *content* on symbols, making them *represent*. But interpretation can't do the same job for visual images, silent sounds, or anything else in our consciousness. And not only are we not doing that when we interpret anything in our consciousness, but we *can't* be doing it. Because if we did, our minds would never finish the job of giving meaning to even one of those images or silent sounds, turning the silent sounds into contentful silent speech.

Suppose you're thinking about, say, Paris, thinking/believing that it's the capital of France. Your belief is about Paris, right? But if it's about Paris the same way a red octagon is about stopping, then it must be because someone or something is interpreting your belief as being about Paris. After all, it's your mind's interpreting the red octagon that makes it about stopping. But for something in your mind to be interpreting (unconsciously) your thought as being about Paris, whatever is doing that bit of interpreting must itself be a thought about Paris, too. Why? Because it has to treat your conscious thought as being about Paris, and that means it has to be about your thought *and* about Paris *and* about the fact that the thought is about the city. We started out trying to explain how your conscious thought that Paris is the capital of France is about Paris, and our explanation is that there is another thought that makes it be about Paris, but that other thought is about three things—Paris, the first thought, and the relation of "being about" that holds between the first things. Well, if this new triple-aboutness thought gets its aboutness in the same way the red octagon gets its stop

2. Why are we trying to answer these questions? Historical narratives explain events in history by uncovering the beliefs and desires of the principals—the movers and shakers. Our confidence that these explanations are at least in the right ballpark stems from our consciousness that our beliefs and desires have content and represent the way things are and the way things could be. So we're exploring how our beliefs and desires do this.

aboutness, there will have to be another thought interpreting the triple-aboutness thought. Let's stop and count up: there will have to be four aboutness thoughts—and so on ad infinitum. That's why we'd never be able to finish giving any conscious thought its content, if conscious thoughts have their content the way red octagons have their content. But if our thoughts, especially our beliefs and desires, don't have their character as representations the way red octagons and yellow triangles do, how do they get that character from our consciousness?

Recall that we are trying to vindicate the idea that there is some fact of the matter about what people actually believe and desire that drives their actions. We may not be able to say exactly what that fact of the matter is, in the case either of the kaiser or even of our closest friends. But there is something that they believe and want, and it does produce their actions. Our confidence in this conclusion comes from our own introspective awareness, from our own consciousness that our thoughts have meaning, content.

Let's look at our own conscious experience carefully, to see whether what seems obvious—that we experience the *aboutness* of our thoughts directly, immediately and unmistakably—is right. We've seen that our thoughts can't get their content by conscious or unconscious interpretation—the way other things, like symbols, get their content.

Let's try on the notion that it's silent speech or mental word images in our consciousness that are intrinsically *about* the things they "name." Well, how could they? Recall the red octagons of street signs. There is nothing intrinsically "stop-ish" about them. If the silent speech in our minds when we think about, say, cows, gets its content just by being a conscious thought, it will have to be intrinsically "cow-ish." But there just isn't anything intrinsically cow-ish about the single silent syllable or the four-letter mental image of the English word "cows."

Perhaps, instead, we think in mental pictures, not in silent speech or mental word images, and that the content of these mental pictures consists in our consciousness of them? Suppose the conscious thought *about* cows is *about* cows because we are conscious of a mental picture that's shaped like a cow. But what makes it a picture of and therefore about cows and not a picture of and therefore about, say, crows? Can it be the shape—mental pictures of cows are shaped like cows, so cow pictures in the mind's eye are about cows? But then it's not our consciousness that gives mental pictures their content; it's their *similarity* to things that makes them be *about* those things. For consciousness to ground the power of thought to *represent*, have *content*, be *about*, it's got to do the work of grounding, not pass it off to some other feature of thought. Consciousness may tell us that a mental picture has a certain shape or smell or sound, but mere consciousness alone can't tell us that it's the shape or smell or sound of a cow. That requires recognition, comparisons, and memory. Or at least it does if what the mind does when it recognizes is the same as what we do when we recognize a painting of a herd of cows as being *of* or *about* cows.

Here is a thought experiment that may help us see why it's tempting to assign consciousness an important role in giving thoughts their content. It also shows clearly what mistake we make when we give in to the temptation. Think about what it's like to sit in a train station waiting room or in an airline terminal in a foreign country, say, Germany, eavesdropping on other people's conversations first in a language you do understand and then in one you don't. And assume for the sake of argument that you don't understand German, not at all. Most of the sounds made by people speaking a language you understand (English) and those made by people speaking a language you don't (here, German) are acoustically the same. For example, you'll hear an English speaker say "nine" and a German speaker say "nein" (the German word for "no"), which, again for the sake of argument, you don't know. They sound exactly the same. As psychologists and philosophers describe it, the sounds of the English and German words are "phenomenologically identical"—the sheer qualitative features of the sound experience are the same when you hear English "nine" and when you hear German "nein." But the phenomenology of the conscious experience in which these two identical sensations are embedded will be quite different between eavesdropping on English speakers and being distracted by German ones. In the first case, you have the conscious experience of understanding what the speakers are saying; in the second, you hear just sounds without meaning. The sounds of English words that you're conscious of have *content*, are *about* something, *represent* the way the speakers think, the way they believe the world is arranged or the way they want it arranged. The sounds of the German words that you're conscious of have none of these things, at least none for you. What's the difference? The difference is obviously in the ways in which the English speakers' sounds cause other images to enter your consciousness, compared with the images, if any, evoked by the German sounds. That's all the phenomenology of the two

experiences consists in. The apparent *aboutness* or *content* that consciousness contributes consists in differences in the images that cross your consciousness immediately after the English ones, compared to the images, if
any, that cross your mind after the German ones. When we think carefully
about this phenomenology, it's pretty clear that the *aboutness* of the English
sounds consists in their being followed in consciousness by other images
and silent sounds. But these new images get their *aboutness,* their *content* by
giving rise to still other images. Could that be all that the feeling of aboutness of thoughts consists in?

Here's another way of convincing yourself that the feeling of aboutness
in your consciousness of the images and silent sounds of your thoughts is
just more images and silent sounds. Suppose your mind's ear "hears" the
following sequence of silent sounds: "dogs dogs dog dog dogs." Mean anything? Probably not. But it doesn't take much for you as a native or fluent
English speaker who knows that "dog" is both a noun and a verb to turn
these five groups of silent phonemes into the consciousness of a full, meaningful English sentence. All that is required is is for your mind's ear to hear
them with the intonation such a sentence would normally receive. Now
your hearing the silent sounds "dogs dogs dog dog dogs" gives rise to a
procession of images in your consciousness of dogs nipping at the heels of
other dogs (dogging them) while they nip at the heels of still other dogs. The
conscious *aboutness of* one set of images and silent sounds just consists in its
being associated with another set of images and silent sounds. What makes
one set of silent sounds into silent speech is just a matter of what further
silent sounds follow the first set.

So, the phenomenology of our conscious experiences is just not going
to help us figure out how our belief that, say, Napoléon was born in 1769
is *about* Napoléon, *contains* a sentence about his birthday, and *represents* the
date in question. Indeed, our conscious experiences are so good at masquerading as the source of *content* for our beliefs that we're going to have to keep
reminding ourselves that they are in the chapters to come.

Let's get back to the kaiser and what was going through his consciousness when he finally directed his foreign office to send the fateful "blank
check" telegram to the government of Austria-Hungary. There was presumably a sequence of silent sounds in German since the kaiser thought in
German. Those sounds had a series of associations, which followed and
were followed by other silent sounds, and perhaps by vague mental images,
too. What exactly were the silent German sounds about? What was their

content? What exactly did they mean? Did they mean or were they about the silent sounds and mental images that followed them or were followed by them in his mind? No, if the kaiser was the authority on what he was thinking *about*, on the *content* of the beliefs and desires that led to his decision, it wasn't the stream of mental "sights and sounds" in his consciousness that gave him that authority.

If our thoughts have content, as the theory of mind demands, it's not consciousness that confers this content upon them. It must come from somewhere else. Consciousness doesn't guarantee we have thoughts with content—beliefs about the way things are and desires about the way we'd like them to be. Indeed, all it seems to reveal is that sensory data are followed by more sensory data. If the kaiser's thoughts had content, it wasn't his conscious awareness of them that gave them their content, as the thought experiments over the last few pages have shown. If we're to vindicate the theory of mind and, through it, the validity of historical explanation through narrative history, we'll have to look elsewhere for what gives thoughts their content.

But where can we look?

Because consciousness is a brain state, a matter of neurons transmitting electrical currents, the *aboutness* of thoughts must be a matter of the neurology of how consciousness comes about or how it works in the brain. Although we don't know much about the cognitive neuroscience of consciousness, it's pretty clear from the "dogs dogs dog dog dogs" thought experiment that if conscious thinking communicates thoughts with content, they must already have that *content* by the time they get to consciousness.

One mainline approach to consciousness begins with the notion that it's the brain's way of focusing its attention. According to this approach, consciousness takes place in a "global workspace" (Baars, 1997; Prinz, 2012). All the modules of the brain, each devoted to a distinct cognitive process, operate in parallel, that is, at the same time. These modules are all controlling and monitoring various parts of the body without continual conscious awareness of all this monitoring and controlling. These distinct unconscious cognitive processes compete to temporarily occupy this global workspace. Among them are the outputs of perceptual regions of the brain, the solutions of problem-solving modules, the bodily movements "decided" upon by brain modules devoted to planning or the "decoding" and "encoding" of speech. These brain modules operate in parallel. Whichever module gains temporary access to the global workspace broadcasts its information content to the other modules, presumably through its presence in conscious awareness. Thus, in effect, conscious attention consists in a brain module accessing the workspace.

In fact, a good deal of empirical evidence supports the finding that there's no difference between the way our conscious thoughts operate in us to explain and predict the behavior of other people and the way they operate to explain and predict our own behavior. This is what we'd expect if consciousness were just a channel of attention and not itself a source of thoughts, inferences, calculations, interpretations, and so on. What neuroscientific and clinical data about normal and mentally ill people show is that what we know about our own beliefs and desires has exactly the same source as what we know about other people's beliefs and desires (Carruthers, 2015). In fact, to explain and predict our own behavior, we turn the theory of mind used to explain and predict the behavior of others on ourselves. And yet, in spite of this, we think we have some special, direct access to what we really believe, unlike our indirect, fallible, and incomplete access to what other people believe.

Functional magnetic resonance imaging and transcranial magnetic stimulation both reveal that the same regions of the brain are active in thinking both about our own actions (what we're about to do) and about the actions of others (what they're about to do). More important, in several psychological disorders—most notably, severe depression and schizophrenia—patients' inability to deploy the theory of mind to understand other people's actions goes along with their inability to explain their own actions, even to themselves. If these two capacities were distinct, we should expect to find one without the other, at least sometimes. This doesn't seem to happen. It turns out that when we look into ourselves, we're using the same evidence and applying the same theory as when we try to figure out what other people are doing. The big difference between figuring out what we're about to do and what other people are doing or about to do isn't the kind of evidence we have to work on. In both cases, the evidence is sensory experience—sounds and sights, smells and tastes (plus, for ourselves, bodily feelings). The only difference is the amount of evidence we have for ourselves, especially the evidence of silent speech and in some cases mental images. There is vastly more of it, but it's just as fallible as the evidence we use to figure out what other people believe and desire. When we're trying to understand other people's actions, we use our sensory experiences of their surroundings and make guesses about their immediate aims. Thinking through what we ourselves are about to do, the content of our conscious processes of thought consists in the same: sensory images, silent speech, and the sensory data that shape both of these. Whether we apply the theory of mind to ourselves or to others at any one time is a matter of where our attention and our consciousness are directed (Carruthers, 2015).

The global workspace model has much to recommend it as a working theory of the functional or causal role of consciousness. There's increasing neurological evidence, including a good deal of fMRI neuroimaging data, for the theory that the neural circuitry of the brain has something like this architecture (Dehaene and Naccache, 2001; Baars, 2002). But the global workspace theory shares with most competing theories of consciousness an acceptance that the *aboutness* or *content* of thought is not produced in consciousness any more than what is illuminated by a flashlight is created by the flashlight's illumination. If there is *aboutness* in consciousness, it wasn't created there, but imported from somewhere else. If we're to figure out how our beliefs and desires have any content at all, let alone the content that, according to the theory of mind, brings about actions, consciousness just isn't going to help.

We are conscious of narrating our lives to ourselves. We are conscious of narrating our lives to others, and we are conscious of others' narration of their lives to us. But consciousness makes no contribution to the explanatory power of the narratives. If our narratives have any explanatory power, it will have to come from a vindication of the theory of mind somewhere else. We need to seek that vindication in the way our minds work, in the way the brain implements—consciously or not—the processes that the theory of mind describes.

There is, however, one feature of conscious experience that is highly relevant to the power of narrative over us and the hold of the theory of mind on our lives. Narratives move us. In fact, they move entire nations. Think of the consequences of narratives like *Uncle Tom's Cabin* or the world-shaking consequences of *The Gulag Archipelago*. Reading or listening to a good narrative creates emotions, at the very least the emotion of satisfaction when it satisfies our feelings of curiosity. This pleasurable emotion is to the theory of mind what orgasm is to sex: two things that were selected for co-occurrence because the pairing increased fitness. Indeed, as noted in chapter 5, good sex and good stories both release the same hormone, oxytocin. But it's not just the satisfaction of curiosity or the pleasures of gossip that theory-of-mind-driven narratives produce. They also generate much stronger emotions, emotions that motivate action. And, of course, we are conscious of our emotions. We feel them and presumably it's our feelings that move us to action. It's this impact on affect that histories like Aleksandr Solzhenitsyn's *The Gulag Archipelago* share with great works of narrative fiction like Harriet Beecher Stowe's *Uncle Tom's Cabin*.

It's because narratives call forth emotions and because emotions motivate us that historians are still arguing about what he had in mind when the kaiser began the war. Even just writing that down—"when the kaiser began the war"—will still raise emotions in the consciousness of some people, people who resent or regret or otherwise feel bad about blaming the kaiser for the war. To this day, more than a hundred years later, the felt emotions of those who blame the Germans for the First World War and those who exculpate the Germans are what makes the question of what the kaiser believed and desired in early August of 1914 a matter of continuing dispute.

We began this chapter with a list of alternative explanations of why Wilhelm started the First World War. More specifically, our problem was to figure out which one or combination of the four belief-desire pairings, if any, that historians have identified was the real cause of his sending the "blank check" to Austria-Hungary. Most people agree that's what triggered the war, more than a month after the assassination of the archduke of Austria. Of course, as historians acknowledge, we'll never conclusively know the kaiser's motives because the evidence of them has been lost or never existed. What we would need to do is to get inside the kaiser's mind at that time, which is too much to ask. But there had to have been definite or indefinite thoughts in his mind with content that resulted in his action. That's what separates history from historical novels. There has to have been a fact of the matter about Wilhelm for the theory of mind to latch on to. Surely, that much we can have absolute confidence in? After all, we know from our own conscious experience that there are facts of the matter about what we believe and desire and that they explain what we do. It was at this point that we had to raise the question of what exactly consciousness really does tell us about our beliefs and desires and how they explain our actions. What became clear in answering that question is that, if we're to vindicate the validity of narrative historical explanations, we'll need much more than our conscious awareness. We'll need neuroscience.

7 Can Neuroscience Tell Us What Talleyrand Meant?

Charles Maurice de Talleyrand-Périgord was a diplomat who successively served Louis XVI, revolutionary France, then Napoléon, and finally the restored Bourbon kings, Charles XV and Louis-Philippe. Only the subtlest and most devious of foreign ministers could have survived, let alone thrived over the fifty years he was at the pinnacle of European diplomacy. He was so devious that when he died, the Austrian foreign minister, Count Metternich, is famously (but falsely) reputed to have said, "I wonder what he *meant* by that?"

The point of the anecdote is obvious. We're all forever trying to read other people's beliefs and desires from their actions. This is much harder in some cases than in others. All his life, Talleyrand tried to make it impossible to figure out what he really wanted and what he thought might be the best way to attain it. Everything he did seemed to be a ploy or to reflect some stratagem no one could fathom.

The job of narrative history is to uncover the motives and designs that explain human actions. The persistent fascination with what Talleyrand did mean, and not just what he could have meant, by the many stratagems he seemed endlessly to deploy, has made him a repeated subject of the historian's and especially the biographer's art. The dozens of biographies written about Talleyrand include one by First Lord of the Admiralty Duff Cooper (Cooper, 2001 [1932]), an important British cabinet minister before World War Two, and another by Crane Brinton (Brinton, 1963 [1936]), great Harvard historian who shaped a generation of scholars. Cooper and Brinton had, between them, much to figure out about their clever quarry: why an archbishop would leave the Church to support the French Revolution, why Napoléon's foreign minister would betray him to the emperor's

Austrian and Russian opponents, why Talleyrand was allowed to remain at the French court even after Napoléon discovered his treachery, how he managed to shift allegiance to the restored Bourbon king, Louis XVIII, and, finally, how he managed to subvert the Congress of Vienna, restoring defeated France to the first rank of European powers.[1] This last achievement drew the attention of a young American scholar, Henry Kissinger.

Applying himself to the matter, Kissinger came to believe he could discern what Talleyrand had in mind, what his beliefs and desires were, at least the ones that explained his decisions and choices at the Congress of Vienna (Kissinger, 1956, 1957). He wrote his doctoral dissertation and many books on the subject. In his conduct of American foreign policy under Presidents Nixon and Ford, Kissinger would be guided by what he thought he'd learned about Talleyrand's (and Metternich's) beliefs and desires, employing the theory of mind. (We'll consider the lessons Kissinger drew from his historical narrative of Talleyrand's achievements at the Congress of Vienna in chapter 11.)

But how could Kissinger have been so sure he knew what Talleyrand was thinking? All he had to go on was what Talleyrand did and wrote (which included a vast preserved correspondence)—and these are the very things

---

1. Taking at face value what Talleyrand wrote down in his letters and diplomatic aides-mémoire, Cooper and Brinton didn't disagree about much, but on what was going on in Talleyrand's mind, they certainly diverged, at least in some instances. In 1809, Talleyrand allowed himself to be seen in Paris with an old enemy, Napoléon's Minister of Police Joseph Fouché. Brinton writes: "Clearly two such Machiavellian characters had not come together for mere love.... Was it to restore the Directory? The Bourbons? Was it to put Murat on the throne? It is more likely that both Talleyrand and Fouché were afraid that [Napoléon's war in Spain] ... might end in complete disaster, and that they wished to seem to have deserted the Emperor in time to act as *king makers*" (Brinton, 1963 [1936], p. 153; emphasis added). Based on the same sources, Cooper comes to quite a different conclusion about what was going on in Talleyrand's mind, writing that Talleyrand "knew his words and deeds [being seen in private conversation with Fouché] would be reported and that Napoléon could put only the worst interpretation on them. The explanation can only be that it was his policy at the time to form the nucleus of an open opposition which ... might thus become strong enough *without overthrowing Napoléon*, to exercise so powerful an influence as to compel him to alter his policy in the direction in which all moderate men desired" (Cooper, 2001 [1932], p. 184; emphasis added). Who's right? Cooper? Brinton? Neither? There has to be a fact of the matter here, doesn't there, even if we can't ever establish what it was?

that beliefs and desires are supposed to explain. How can historians avoid the dangers of circular reasoning, which deludes them into thinking they understand, but which leads them to write what amounts to historical fiction, stories that do no more than scratch the itch of our curiosity?

You'd think at least one of the possible stories of what Talleyrand believed and desired must be right. Perhaps Kissinger and, indeed, every other historian until now got Talleyrand wrong. But, as we came to see in chapter 6, there has to be a fact of the matter about what Talleyrand thought. Without it, narrative history can't explain the actions of anyone. Without it, narrative history is not just wrong; it's impossible.

The point here is not that narrative history is fallible. We know that applying the theory of mind to the available evidence is never enough to be completely confident that our historical or biographical explanations are right. That's one reason why these explanations constantly get rewritten. The available evidence doesn't rule out alternative hypotheses about the beliefs and desires that drive the actions we want to explain. The evidence we need to do that is supposed to be in people's minds, and we just can't get into their minds. All that is true enough. But the problem is much deeper.

For narrative history to even stand a chance of truly explaining people's actions, we have to be confident that there really are beliefs and desires in people's minds and that they really do cause them to act. Chapter 6 showed that our own conscious experience of our thoughts provides us no assurance that there is a fact of the matter about what we believe and desire, and thus no basis for confidence that there is such a fact of the matter about what other people believe and desire as well. If consciousness can't provide the assurance that our minds work the way the theory of mind tells us they do, what can?

There is one, and only one, recourse, one path out of this impasse. We know that the specific beliefs and actual desires about which narrative history can only hypothesize are to be found in people's minds. Unless the mind is a distinct entity from the brain, surely, the beliefs and the desires that bring about behavior are somehow inscribed in the brain, in the neural circuits of the cerebral cortex.

If the mind and the brain are in fact not distinct but one and the same thing, then our confidence in the theory of mind as the right theory to explain human thoughts and actions gives us the very tools we need to solve narrative history's problem, to figure out, at least in principle, what people think. Of course, an "in principle" solution is all we'll get for the foreseeable future, at least until we can read people's thoughts from their brains in real time. But because, in most cases of historical inquiry, there will be no pressing need to read the exact beliefs and desires that drive an action, a method of doing so will provide at least in principle the ground rules for settling historical disputes. And it will ensure that there is a fact of the matter, a right answer to the explanatory questions narrative history seeks to answer, even if we don't yet have the means to nail that answer down exactly. The only way to vindicate narrative history's recourse to the theory of mind is through neuroscience.

The theory of mind's relevance to neuroscience has been evident to neuroscientists almost since the emergence of their discipline in the nineteenth century. Along with everyone else, neuroscientists have embraced the theory of mind, at least as a first approximation to explaining human thoughts and actions. The theory of mind told them that, to discover how beliefs and desires worked together to generate the body's actions—speech, movements, activities, responses to stimuli—they first had to find where in the brain the beliefs and desires were lodged. The theory of mind established both the initial research program of neuroscientists and the basis for their hopes to vindicate the theory with the successes they anticipated in pursuing that program.

In this chapter, we'll try to find that vindication. We won't look to neuroscience to tell historians exactly what was in Talleyrand's mind. That would, of course, require a time machine and mind-reading equipment, neither of which exists (nor, in the case of the time machine, is ever likely to). What narrative history needs is some assurance that when people act, there really is a desire box and a belief box somehow and somewhere in their brains that brings about their actions. Neuroscientists would vindicate narrative history's explanations if they showed that clear signs of beliefs and desires could be detected in the brain and that beliefs and desires worked roughly the way the theory of mind and thus also narrative history said they did. That may sound easy, until we realize exactly what task neuroscientists had set for themselves. It was showing how the neurons in brain tissue could inscribe, store, and transmit thoughts—beliefs and desires—*about* the world, statements with *content*, how neural circuits could *represent* the ways things are arranged and the ways we want things to be arranged. The task for neuroscientists was thus to show how and why the theory of mind was true, or at least true to a first approximation.

Expecting the theory of mind to drive the research program of neuroscience exemplifies a familiar recipe for success in science. Consider the history of genetics: Gregor Mendel noticed some regularities in the heredity of obvious traits of pea plants—their height, the color of their seeds, whether the seeds were wrinkled or smooth. In 1866, he framed a theory that consisted of two "laws" to explain his observations (roughly the laws of independent assortment and of segregation). The theory hypothesized the existence of unobservable entities (the word "genes" would not be coined to describe them for another forty years or so). Accepting Mendel's theory motivated scientists to launch a research program to identify specific genes for specific traits, to locate these genes in the body, to uncover their structure and composition, their mode of action, how changes in them effect hereditary traits, and so on. The research program driven by Mendel's theory resulted in many Nobel Prizes for the twentieth-century scientists who answered these questions.

Neuroscientists should expect the theory of mind to play the same role in their research program: through the mechanism sketched out in the chapter 3 diagram (figure 3.2), the theory advances hypotheses about the decision-making process that brings about people's actions. It falls to the neuroscientists to fill in the details of this process.

Here again is that diagram (figure 7.1).

Thus neuroscience is tasked by the theory of mind to try to uncover the neural mechanisms that constituted each of the boxes—square and hexagonal—in the figure 7.1 diagram. And then it has to elucidate the mechanisms that underlie each of the arrows in the diagram as well. Naturally, when neuroscientists begin to do so, they might have to change the configuration of the diagram considerably, adding new boxes and new arrows, dividing boxes, or even eliminating boxes altogether in favor of a flowchart quite different from the diagram they start with. In doing so, however, neuroscientists would still be following the lead of the twentieth-century molecular biologists who wrought radical changes to Mendel's theory of heredity, even though the result was recognizably a development and enrichment of the theory he propounded in 1866.

Neuroscientists' exploration of how the brain worked began to take serious scientific shape at the end of the nineteenth century. In addition to the initial guidance the theory of mind provided them, neuroscientists had guidance from case studies of patients with mental illness or conditions, where the theory of mind appeared to break down, where patients behaved

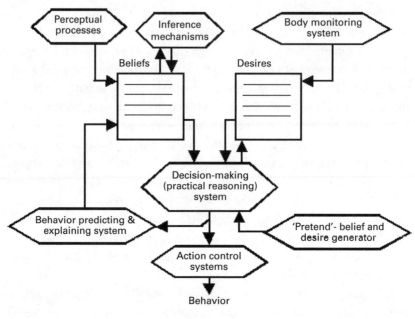

Figure 7.1
"Boxology" of the theory of mind sets out marching orders for the research program
of neuroscience. From Nichols et al., 1996, fig. 1.

in ways contrary to what the theory led neuroscientists to expect. Starting
with such behaviors, the neurologists among them sought the lesions, injuries, defects in the brain that caused them. Adopting a top-down strategy,[2]
they watched how patients with mental illnesses or conditions responded
to stimuli and then sought the brain injuries that were impairing the
patients' reasoning, that were giving rise to their mistaken beliefs and pathological desires, and that, separately or together, were producing "deviant"
behaviors. Psychoanalysis of all types revealed the neurologists' reliance
on the theory of mind when they began to theorize about unconscious

---

2. Although the study of cell physiology might, in principle, have been a productive
bottom-up strategy for neuroscience's research program—to first identify the individual nerve cells (neurons), their networks, and the electrochemistry of their activity
as these related to theory of mind—in practice, it was not; hardly any neuroscientists
adopted it. With good reason: there are some 87 billion neurons in the human brain,
and the number involved in even the simplest mental activity is astronomical.

beliefs and desires. Sigmund Freud was clear in his insistence that such theorizing could only be vindicated by locating these unconscious beliefs and desires in the brain (Kitcher, 1992). Freud's confidence in the relevance of neuroscience to understanding the mind was increasingly vindicated in the course of the twentieth century. Starting with the lesions uncovered mainly in autopsies of cadavers, neuroscientists moved on, first to electroencephalography (EEG), then to neuroimaging (MRI and fMRI), and later to direct electrical stimulation of distinct parts in the brain (TMS). Over the past century, they learned a great deal about the specific location of various sensory, cognitive, and affective mental phenomena and thus also about how the theory of mind appeared to be hardwired into the temporoparietal junction, the dorsolateral prefrontal, the inferior frontal gyrus, and the ventral medial prefrontal cortex (as we saw in chapter 5).

But they needed to locate something like beliefs and desires in people's brains and to figure out how the neural circuits that carried these beliefs and desires caused people's behaviors. Two questions immediately arose. What should neuroscientists look for? And what if neuroscientists couldn't locate something like beliefs and desires in the brain?

The answer to the second question is obvious: if they couldn't locate something somewhere in the brain corresponding at least roughly to the boxes and arrows in the figure 7.1 diagram, the theory of mind would be in big trouble. Although just how big that trouble might be is the subject of chapter 8), to start with, if neuroscientists failed to vindicate the theory of mind as even a first approximation to the truth about how the brain worked, they'd either have to give up the theory of mind altogether or give up the notion that the mind was, even to a first approximation, the brain. Many nonscientists might not be troubled by this second possibility. But most scientists would be deeply troubled. What's worse, wedded as it was to the theory of mind, narrative history would have to admit that its explanations required the operation of nonphysical, nonmaterial processes that couldn't be scientifically studied at all.

The first question, "What should neuroscientists look for?" is one explicitly addressed by the theory of mind. It tells them to find where in brain beliefs and desires are inscribed, stored, and transmitted to guide behavior. In other words, they have to find where the neural circuits that carried the beliefs and desires are located—which networks of neurons, anatomical regions, or modules of the brain encoded them. Then it tells them they to explain, for example, how the neural circuit that carries the belief that Paris is the capital of France differs from the neural circuit that carries the belief that Berlin is the capital of Germany; and to explain what differences each neural circuit makes to the behavior of a person who holds one of these beliefs when the neural circuit is paired with the neural circuit that carries the desire to go to the French capital versus the neural circuit that carries the desire to go the German capital.

For the moment, let's just focus on beliefs, starting with a relatively simple one, the belief people have that Paris is the capital of France. Once neuroscientists figure out how the brain stored that belief, they'd begin to get a handle on how the brain stored more complicated thoughts, like the belief that a war in the Balkans would break up the Triple Alliance.

What the neuroscientists are trying to narrow down, locate, and identify are the neural circuits somewhere in one or perhaps many different parts of the brain that *represent* the fact that Paris as the capital of France, have the statement to that effect for its *content*, and is *about* Paris. Why? Because that's what beliefs are—*representations* of how the world is or isn't arranged; they do that by *containing* statements *about* things, in this case *about* Paris. If something is not a representation, then it's not a belief. So, pursuing the research program dictated by the theory of mind, neuroscientists need to look for how the gray matter represents things.

In particular, what they are looking for in our example is a network of connected neurons that "contain" the statement that Paris is the capital of France, a network organized in such a way that it is "about" at least three distinct things: (1) "about" Paris; (2) "about" its being a capital; and (3) "about" France. This would be a network of neurons—perhaps hundreds of thousands or even millions of them—wired up distinctively and differently from another network of neurons, say, one that "contains" the statement that Berlin is the capital of Germany.

Before the neuroscientists even begin, however, we can rule out some alternative ways in which a network of neurons could *represent*. Could the network represent Paris the same way a postcard or a painting does? A picture postcard and a painting are *about* Paris and *contain* information about Paris presumably because they *look like* some part of Paris. They represent by being "representational," that is, there is some one-to-one relationship between the lines, planes, curves, and colors on the picture postcard or painting and the features of the part of Paris they depict. Of course, Paris is three-dimensional, made of asphalt and marble, bricks and glass, and the postcard and painting aren't. Still, the ratio between the size of images in the postcard or realistic painting and the size of the things in the actual place from the photographer's or painter's perspective would fall within a fairly narrow range of values. That's enough for a sort of "scale model" representation in this case.

Could the neural circuit in a person's brain that represented the fact that Paris is the capital of France be *about* Paris because its "wiring diagram" looked like Paris did in the picture postcard or painting? Obviously not. The synaptic connections in the belief neural circuit that represented the fact that Paris was the capital of France couldn't look anything like Paris because there'd just be no way they could physically "look like" that fact. Nothing in the brain could look like a particular fact in the way representational pictures look like particular things. There are indefinitely many facts about Paris we could come to learn from looking at a photo of it. Which of these facts, once in our minds, would the photo look like? None of them? All of them? If there is a belief neural circuit that stored a fact about Paris, it couldn't do it the way a photo stored facts, even if a neural circuit could look like Paris.

Well, why couldn't a neural circuit record the fact that Paris is the capital of France the way that the printed letters on a page do? Suppose somewhere in the brain there was a neural circuit that was the brain's way of "writing down" and "storing" the fact that Paris is the capital of France. Such a neural circuit would be *about* Paris.

Here an argument from chapters 3 and 6 becomes crucial again. Recall our discussion of what makes the red octagons of street signs into representations of the command to stop. There is nothing intrinsically "stop-ish" about the shape or color or combination of shape and color of these signs. The red octagons represent the command to stop because something inside our minds/brains interprets them that way. For a Paris neural circuit to represent the fact that Paris is the capital of France in the same way (i.e., symbolizing the fact), there would have to be something else in our brain, some additional neural circuit to interpret the Paris neural circuit as doing so. And this further interpreting device in the brain would itself have to represent both (1) the Paris neural circuit and (2) Paris itself and to interpret the first as representing the second, just as our minds/brains do when they interpret the red octagon of a street sign as representing the command to stop. The

Figure 7.2
Road sign showing direction to Paris.

second neural circuit that interpreted the first would have to be both *about* Paris and *about* the first neural circuit. But this would *double* our problem instead of solving it. We would have gone from the problem of how one neural circuit could be *about* Paris to the problem of how a second neural circuit could be both *about* Paris and *about* the first neural circuit.

An analogy might make this problem clearer. Suppose you were trying to get to Paris, but you got lost. Then you saw a rectangle with some white marks on black (figure 7.2).

How would you know that the white marks in the rectangle represent the fact that Paris was to the right of where you were now? Well, your brain immediately and unconsciously would *interpret* the configuration of white and black on the rectangle as being *about* the place you wanted to go. But to interpret the white marks on black as representing Paris, you'd need to have stored information in your brain that already represented Paris, and then to compare that stored information with the sign's white marks on black. The sign's white marks on black would mean the direction to Paris only because your brain *recognized* the marks as being *about* Paris. You'd have a thought *about* the sign, that it was *about* Paris. Knowing that the sign gave the direction to Paris would require that there be two representations in your brain—one of the sign and another of the city that the sign pointed you toward. There would be two instances of *aboutness* in your thoughts: one about the sign, and another about where it was pointing.

Now suppose the neural circuit that represented your belief that Paris was to the right did so in the same way the octagonal road sign represents the command to stop. Or suppose there was a microscopic version of the road sign in a neural circuit somewhere in your brain. For it to represent the direction to Paris, there would have to be another representational neural circuit else-where in your brain to interpret the microscopic road sign. A microscopic road sign in a neural circuit of the brain, no matter what "letters" it was written in, would require that there be another neural circuit to read it, one representing both the rectangular road sign in the first neural circuit and what the sign itself represented—Paris. This would begin a regress, requiring a third neural circuit to explain how the second circuit represented what it read off the microscopic sign in the first circuit, and so on *ad infinitum.*

But sentences aren't the only things that convey information. Maps represent the way things are arranged, too. Could a neural circuit in your brain represent the fact that Paris is the capital of France the way that a map does?

Consider the big five-pointed star in the circle next to the word "PARIS" in capital letters on the map of France in figure 7.3; it indicates that Paris is the capital of France. But how exactly does it do that? Well, most maps come with a "key" or "legend"—a list of symbols usually on the bottom left

Figure 7.3
Map of France represents that Paris is the national capital.

side of the map. If this were a typical map of Europe, in its legend or key, the words "national capital" would be written in capital letters of the same font as "PARIS" and next to it a five-pointed star in a circle. And you'd know the conventions of map displays well enough to use that information to *interpret* this particular map.

So, if your original neural network represented the fact that Paris is the capital of France the way a map does, there would have to be something somewhere else in your brain like a "key" or "legend"—a list of symbols together with some writing that guided the *interpretation* of the symbols.

Do the neurons of your brain work like that? Well, they could, but only if there were some other neural network in your brain that played the role of the key. And there would have to be still another neural network that could interpret the key or legend. But one part of your brain having to interpret another part is just the problem we were trying to avoid. We're back to the original problem we thought the map metaphor might solve: how do neural circuits in the brain represent—in this case how do they interpret the neurons that give the key as symbols, meaning, for example "National capitals are indicated by a five-pointed star in a circle."

We've excluded three amateurish hypotheses about how neural circuits might carry information. We went over their mistakes carefully to be on our guard, as neuroscientists have to be on theirs, against making the same mistakes in considering other, more sophisticated, better-informed hypotheses about how beliefs might be composed of neural circuits.

One obvious thing about beliefs that helps neuroscientists in their research program is that they are held in memory. Most of our beliefs are not present before our minds in consciousness. Thus, for example, we believe that Paris is the capital of France even when we're not rehearsing that thought in our consciousness. Knowing that, a good place to look for beliefs would be in the part of the brain that stores memories, in particular, whatever part stored what cognitive neuroscientists call "declarative" or "explicit" memories—those we describe in sentences we believe to be true.

Here some lessons from clinical medicine became important. A male patient most often referred to only as "H.M." (perhaps the most famous patient in neuroscience) suffered from severe epilepsy. To control it, he was subjected to surgical removal of his hippocampus and several other nearby structures of the temporal lobe, including the entorhinal cortices, in the

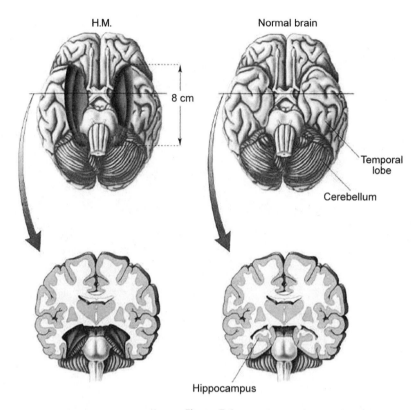

Figure 7.4

H.M.'s brain compared with a normal brain, showing area excised, including the hippocampus and entorhinal cortex. From https://i0.wp.com/i61.photobucket.com /albums/h53/mocost/HM.jpg

1950s. (Much of H.M.'s history and the findings work with him generated are recorded in Dittrich, 2016.)

As a result of the removal of these parts of his brain (figure 7.4), H.M. immediately lost the ability to remember almost all new information, though he retained other abilities, among them the ability to learn new motor skills. This enabled neuroscientists to identify the crucial role of the hippocampus in memory formation. Thus began continuing research into understanding how regions of the hippocampus create, store, and deploy "explicit memories," ones we usually describe as "beliefs expressed in statements" such as the belief that Paris is the capital of France. The most obvious question facing the researchers was, Exactly how are explicit

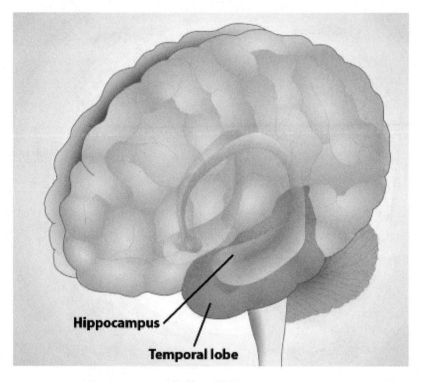

Figure 7.5
The hippocampus, where information is first recorded and stored. From https://sites
.duke.edu/apep/files/2016/02/module-03-figure-02.jpg

memories encoded and first stored in the hippocampus and surrounding temporal lobe (figure 7.5)?

Given what was known about the human brain and its composition in the 1950s, when the surgery on H.M. was performed, that question was also the least answerable one facing neuroscientists. So vast, complicated, and inaccessible to experimenters were the neural networks of the brain that no one really had the slightest idea how to answer this most obvious question. But now, sixty years and three Nobel Prizes later, we're beginning to, and in detail.

The human brain comprises 86 billion neurons, each linked up to a 10,000 or so other neurons. Almost all of these neurons seem mainly to do one special thing, and do it continually: firing in synch, over and over again, in a vast number of input/output circuits, they move discrete electrical charges, "action potentials" to other neurons. They do this largely by moving a few different kinds of neurotransmitters—small charged molecules that can defuse rapidly in the gaps (called "synapses") between neurons. The only neurons that don't work in almost exactly this way are the ones that respond directly to sensory inputs and the ones connected to muscle fibers, although even these connect to other neurons in the same way that most neurons do. When sufficiently many electrical signals are sent from one neuron to another over a short enough period, the neurons build new synapses that reach out to others (Kandel, 2000). The mechanism by which this happens is well understood. The increase in neurotransmitter movement causes a chain of events inside the neuron back to certain genes in its nucleus that switches them on. They start to produce new proteins to build new synaptic connections that make electrical signal transmission easier. This vast number of input/output circuits, which include larger input/output circuits composed of wired-up sets of smaller input/output circuits and are all pretty much the same in their molecular neurobiology, carry all the information the brain stores. If that's all the neural circuits ever do—fire in synch, over and over again—then how they encode, store, and express beliefs stored as explicit memories must be in the details of their firing patterns.

We are going to dive into the details. Like all science the details will be difficult to absorb and to keep in mind. That's because science isn't stories. But the details turn out to be of the greatest importance for the prospect of grounding the narrative explanations of history and biography. In fact they will unravel any confidence we might have had that history is more than just a collection of engrossing but utterly fictional stories. That is why we need to go into the details. It's only by seeing how the brain really works, and what it means for the theory of mind, that we can come to grips with the problems that confront history's claim to explanatory knowledge.

Learning what H.M.'s symptoms revealed, the neuroscientist Eric Kandel set out to answer the question of how the neurons store memories. His work was rewarded by the Nobel Prize in 2000. This work was extended, deepened, developed by the discoveries, first, of John O'Keefe and, then, of May-Britt Moser and Edvard Moser, for which all three won the 2014 Nobel Prize. Between them, Kandel, O'Keefe and the Mosers revealed exactly how the brain encodes the information that the theory of mind says is contained in our beliefs. What Kandel discovered was very troubling for the theory of mind, as we'll see. And then matters were made worse for the theory by what O'Keefe and the Mosers revealed.

Kandel knew from the beginning there was no point in trying to figure out how neural circuits in the human brain stored beliefs in the form of memories. So he and his lab decided to start small, on a simple model system: the brain of the sea slug *Aplysia californica* (figure 7.6), which has a small number of very large neurons. Since the sea slug's brain doesn't store explicit beliefs in its memory, it wasn't at all clear that this research would have anything to say about how the human brain stored them. But it did, as Kandel's Nobel Prize shows.

Even though sea slugs don't acquire new beliefs that their brains can store as explicit memories, they can learn new behaviors by classical conditioning and "remember" them, at least for a while. Recall how Ivan Pavlov conditioned dogs to salivate at the sound of a bell when it had previously been rung at mealtimes. The sea slug can be conditioned too: touch its front and it won't do anything much. Give it an electrical shock and it will

Figure 7.6
Aplysia californica, the sea slug, whose brain served as Kandel's model system. From https://upload.wikimedia.org/wikipedia/commons/e/ef/Aplysia_californica.jpg

shrink back. Touch its front while giving it an electrical shock enough times and eventually it will shrink back when you touch its front without the shock. Kandel found that, through conditioning, it could learn a response and remember it, at least for a while. He called what the sea slug learned an "implicit memory" because the sea slug had acquired a new *ability*, a new disposition or capacity to respond to stimuli. Because its brain had a small number of large neurons, Kandel was able to identify exactly which neurons were involved in storing the implicit memory and exactly how they did it—which anatomical changes, driven by which somatic genes producing which particular proteins, resulted in new neural circuits that encoded the newly learned behavior (Kandel, 2000).

Kandel also found that implicit memories in the sea slug came in two versions: short- and long-term, depending on the number of training trials to which its neural circuits were exposed. A little conditioning—a few front touches plus shock associations—produced a short-term implicit memory, one that wore off after a short time. More conditioning produced an implicit memory that lasted longer. Kandel was able to exploit advances in neurogenomics—the use of gene-knockout and gene-silencing techniques in the study of neurons—to show that the difference between short- and long-term implicit memory was the result of switching on the somatic genes in the neurons that build new synapses.

The difference between short- and long-term implicit memories in the sea slug was reflected in a fairly obvious anatomical difference: a short-term implicit memory appeared to be a matter of establishing temporary bonding relationships between molecules in the synapses, bonds that degraded quickly, whereas a long-term implicit memory appeared to be a matter of building more new synapses between the neurons. The former produced "short-term potentiation" or STP the latter, "long-term potentiation" or LTP. ("Potentiation" is neurospeak for increase in signal transmission at the synapses.)

Short-term implicit learning occurred when a few touches plus shocks provoked a chain of molecular events that briefly modified the number and shape of neurotransmitter molecules in existing synapses that already linked neurons. The modification of the neural pathway to respond to touch stimulus alone lasted for a short time, as the neurotransmitter molecules diffuse and degrade.

Learning to respond that way for a long time—long-term implicit memory through the process of long-term potentiation or LTP—occurs when touches and shocks are paired many more times. The repeated pairings at first produce short-term implicit memory, but keeping this stimulation up produces more neurotransmitter molecules, which then diffused back from the synapses to the neurons' nuclei, where they switch on somatic genes to produce the building blocks of *new* synapses. These new synaptic connections work in the same way that the smaller number of synaptic connections laid down for short-term implicit memory do, but their larger number means that the learned response will continue even if a significant number of the synaptic connections degrade, as they do over time.

Once Kandel and his team had revealed the mechanism of implicit memory in the sea slug, they showed it was pretty much the same mechanism—same neurotransmitters, same somatic genes, same synapses or synaptic connections—that produced implicit memory in other species, such as the roundworm *Caenorhabditis elegans* and the fruit fly *Drosophila melanogaster*.

What they had discovered in the sea slug was nothing less than Pavlov's classical conditioning mechanism. But what does the classical conditioning of sea slug neurons have to do with beliefs, like Paris is the capital of France, being stored by human neurons? Everything, it turns out. When we acquire new beliefs and store them in memory, the neurons in our brains do exactly the same thing that the neurons in the sea slug's brain do when it acquires and stores new behaviors—only with a lot more LTP and lots more neurons growing new synapses.

Our explicit memories are composed and stored in the hippocampus (figure 7.4, right), then moved to (neuroscientists would say "consolidated in") the gray matter of the neocortex that immediately surrounds the hippocampus. Now, these two brain structures are completely absent in the sea slug, roundworm, and fruit fly. Nevertheless Kandel and his colleagues were able to show that the *same* molecular mechanisms and the *same* somatic genes that build new synaptic connections responsible for acquiring and storing implicit long-term memories in the sea slug, roundworm, and fruit fly are also responsible for acquiring and storing explicit long-term memories in vertebrates, mammals, primates—and us (Kitamura et al., 2015). It turns out that our acquisition and storage of explicit memories—beliefs—are just long-term potentiation or LTP.

But, wait, just because the lowest-level neural processes appear to be the same across the phylogenetic spectrum from roundworms, fruit flies, and sea slugs to humans is no reason to say that there's no difference between, say, how the brains of sea slugs acquire and store long-term memories and how ours do. Surely, the vastly greater number of neurons in our than in sea slugs' brains makes a significant difference in itself? Surely, there are other significant differences between our and sea slugs' brains: how neurons are wired together, for example, and whether higher-level structures, anatomically distinct parts, specialized regions, dedicated areas, and modules are present or absent in the brains. Surely, all the beliefs acquired and stored as long-term memories can't just be chalked up to LTP in both sea slug and human brains? This is an objection that readers are likely to raise repeatedly in what follows. We'll only be able to address it after reviewing other findings by neuroscientists. So, bear with me for the moment (we'll come back to this argument at the end of chapter 9).

In Kandel's original experiments, rats were motivated by fear of drowning to acquire an explicit belief: that a hidden platform was at a certain location in the deep-water pool in which they were placed (figure 7.7).

With careful study of the experimental rats' hippocampi (the rat brain has both neocortex and hippocampus) before, during, and after they learned the location of the hidden platform and acquired the belief about its location, Kandel and his team showed that long-term memory of the

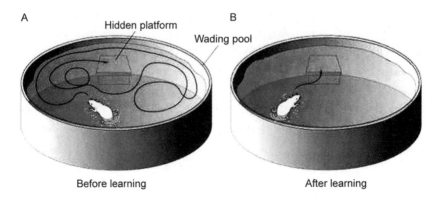

Figure 7.7
Morris water maze, used by Kandel to uncover explicit memory mechanisms in the rat. From https://nwnoggin.org/wp-content/uploads/2015/05/Water-maze.jpg

platform's location is the same process (long-term potentiation or LTP) in the rat, right down to the neurotransmitters and the somatic genes, that it is in the sea slug. They then were able to show that LTP of explicit memories works the same way in humans—same new synapses, grown by the same process of somatic gene regulation in the same part of the brain, the hippocampus (Bailey, Bartsch, and Kandel, 1996).[3]

Other researchers would find that, once acquired by LTP in the hippocampus, explicit memories are moved to (consolidated in) information storage circuits in the neocortex—the visual, auditory, and parietal cortices— by the same molecular biology of LTP. They were able to show that it works by the same molecular modifications of neural circuits in the human brain that Kandel's team discovered in the sea slug brain when it acquired long-term implicit memories—long-term dispositions to respond to stimuli (Kitamura et al., 2015; we'll go more deeply into the relevant details, and why they're relevant later in the chapter).

The details of the neural connections involved in long-term storage of implicit memories (abilities and dispositions to behave) differ only by number from the details of the neural connections involved in long-term storage of explicit memories (beliefs). The difference in number is great, however: in the sea slug, the number of neurons that have to be wired up to store a bit of conditioned behavior might be a few hundred (after all, its brain has only 18,000 neurons in all). In the human brain, the simplest stored belief is going to involve hundreds of thousands of neurons. But whether a few hundred or even a million neurons, what's going on in the brain of a human is the same thing that's going on in the brain of a sea slug; there's just much, much more of it.

Actually, the discovery that the sea slug's neurons and ours do exactly the same thing should come as no surprise. It vindicates an old maxim that goes back to the eighteenth century: "Natura non facit saltum," literally, "Nature does not make a jump." Substantively, differences in the biological domain are matters of degree, not of kind. What Kandel and his colleagues discovered is that abilities stored as explicit memories in the rat brain at least are

3. Bailey, Bartsch, and Kandel were also able to show that explicit memory in the rat works exactly the same way that implicit memory does in the sea slug *Aplysia*, with both short- and long-term potentiation (STP and LTP; Bailey, Bartsch, and Kandel, 1996).

just a much greater number of the same sort of abilities stored as implicit memories in the sea slug brain. They showed that neural circuits didn't store information by *representing* it, being *about* it, having it as their *content*. They showed that storing explicit memories was a matter of neurons being connected to one another to produce certain kinds of results, events inside the brain, in the neural networks, and eventually in behavior that other animals, like humans, could detect. Convincingly demonstrating that what Kandel and his colleagues had reported was in fact true took considerable work, work that earned three other neuroscientists, John O'Keefe and the team of May-Britt and Edvard Moser, a Nobel Prize in 2014.

By themselves, of course, Kandel's findings would probably not convince you of much. The similarities his team found at the level of individual neurons may not be the whole story, or even much of it. When a difference in degree is as great as the difference between LTP in a brain of 18,000 neurons and LTP in a brain of 87 billion, the result might well be a difference in kind. Experimenting on rats and sea slugs won't tell us, nor did it tell neuroscience researchers, how the human brain stores beliefs that can be expressed in sentences like "Paris is the capital of France."

But, not being able to directly experiment on humans, what could neuroscientists who were curious about these matters do? What they needed was, first, animals with brains sufficiently similar to ours that could be experimented on in large numbers and without raising ethical qualms. Sufficient similarity would have to be a matter of having brains with the same parts arranged in the same way, but with a lot fewer neurons, of course, since no nonhuman animals have brains even close to as big as ours except apes, and ethical concerns rule out experimenting on them almost as strongly as they rule out experimenting on humans. The obvious candidate animals were still the ones Kandel employed, rats. Whether the rat brain was sufficiently similar to ours to be a good "model system," couldn't be decided in advance. It would be decided by whether results of experiments on the rat's brain enabled researchers to make precise and reliable predictions about human brains and how they worked. So far, the overwhelming evidence amassed by neuroscientists is that the similarities are great enough at every level of organization to make the rat brain a good, though not perfect, "model system" for the human brain (figure 7.8, plate 2) in the areas of cell physiology and, to a large extent, gross anatomy and physiology, as well as in the areas of clinical medicine, pharmacology, psychopharmacology,

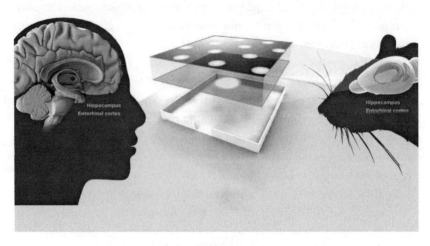

Figure 7.8

Relevant parts of the rat and the human brain—the hippocampus and the entorhinal cortex. From https://www.extremetech.com/wp-content/uploads/2014/10/nobel
-prize-grid-cells-diagram-place-cells-rat-human-640x353.jpg. With permission from
Mattias Karlén/The Nobel Committee for Physiology or Medicine.

and clinical psychology. The great thing about experimental neuroscience
is that hypotheses researchers frame about the human brain on the basis of
rat studies can be tested, at least in principle.

There is, however, an obvious problem in using the rat brain to figure
out how the human brain acquires and stores beliefs. It's much harder to
try to figure out how the rat brain acquires and stores rat beliefs than how
the human brain acquires and stores human beliefs since we don't really
know exactly what rats believe to begin with. We can tell one another what
we believe in spoken language. Rats can't, at least not in any language we
can understand. So, the first problem neuroscientists faced was to find a
set of sentences in English or any other spoken language that described
some statements that rats by their *behaviors* clearly appeared to believe
and then to search for where in the rat brain these beliefs were acquired and
stored, and for how they were acquired, stored and used. But in order to
have much bearing on how our brains acquire, store, and use beliefs, the rat
beliefs would have to be about statements of the same kind we could and
do believe ourselves. Researchers would have to find beliefs that rats clearly
appeared to have that are sufficiently like our belief that Paris is the capital

of France, for example, so that they could with confidence infer from how the rat brain carries and uses its beliefs to how our brains carry and use the belief about Paris.

Are there sentences in English (or Norwegian as we'll see) that express statements that experimenters could confidently attribute as believed by both rats and humans? If there are, experimenters could try to figure out how the rat brain stores and uses these beliefs and then ask whether the human brain does it the same way. What experimenters need to do is find unambiguous rat beliefs that they could reliably read from rat behaviors and then look for where and how these beliefs are stored in the rat brain. Reading rat beliefs from rat behaviors is really not so different from how we read other people's beliefs from their behaviors, mostly from what they say or write. But saying and writing are still behaviors and inferring from them exactly what people believe is often pretty dicey, even when they're sincere and speak our language. Sometimes it's better to ignore what people say and instead watch what they do. In some respects, then, what rats do—their nonverbal behaviors—may be as good a guide to what rats believe as what we say or write is to what we believe—or even a better guide.

Here are some good candidates for unambiguous beliefs we and rats share that are sufficiently like the belief that Paris is the capital of France: beliefs about our or their current location, about the path we or they took to get there, about the direction from which we or they traveled and how fast or slow, the obstacles we or they had to circumvent, about where food, water, warmth, electrical shocks are, about what choices we or they face in the near future when we or they search for food, water, warmth, and so on. These are shared beliefs that experimenters can pretty safely read from rats' behaviors, especially after they have trained up the rats in their labs.

Figuring out how the rat brain encodes, stores, and uses beliefs from rat behaviors won an Irish American New Yorker transplanted to Britain, John O'Keefe, and two Norwegians, May-Britt and Edvard Moser, a Nobel Prize in 2014, fifteen years after Eric Kandel won his. What O'Keefe and the Mosers discovered was how the rat brain and the human brain store beliefs. And, as we'll see, what they also showed was that *nothing* in the rat's brain or in ours works *anything like* the way the theory of mind says beliefs and desires work—as representations with content expressed in statements about things.

O'Keefe and his coworkers were focused on the hippocampus because, like Kandel, they knew from amnesia studies in humans (especially in patient H.M.) that it forms and initially stores beliefs we often express as explicit memories. What they discovered were the first indications of exactly how the neural circuits in the hippocampus do this. Actually, it was by accident that O'Keefe uncovered what he called "place cells" in the hippocampus. These cells are the exact location in the brain where information about the current position and trajectory of the body is encoded (O'Keefe, 2014). Twenty years later, the Mosers discovered cells nearby that store information about the geography of local environments, its boundaries, and the body's orientation and speed. They called these cells "grid cells" for reasons that will become apparent. The Mosers also figured out how the information that grid cells store is fed to the place cells (Moser, E., 2014, Moser, M-B., 2014). Within a few years after their work became known, the role of these cells in recording other nongeographical information was vastly expanded (Manns and Eichenbaum, 2009). Eventually, the techniques O'Keefe and the Mosers employed were used to identify other neural circuits dedicated to recording the full range of information about the rat's past, present, and future environment—its beliefs about its world. Making our way through some of these discoveries, we'll see how differently the rat brain and the human brain work from the way the theory of mind requires them to.

It will be important for the lessons to be extracted from this research that we begin by explicitly adopting the perspective of the experimenter and not the subject. We need to learn what they discovered about the hippocampus and the entorhinal cortex next to it (figure 7.9), and then ask how these two brain regions work together to store and use information.

Wire up individual neurons in a rat's brain and put it in a square or rectangular or round box. Electrodes positioned over different neurons in the medial entorhinal cortex will fire whenever the rat passes a particular spot in the box (figure 7.10). Give each neuron that fires a label—a number, a letter, whatever. Let a rat walk around long enough and you can identify where it is in the box just by seeing which neuron in its brain fires. Mark the floor of the box with the number of the neuron that fires when the rat passes that spot.

What you get when you run this experiment is a remarkable grid across the whole floor of the cage, not an x,y square grid, but a grid composed of hexagonal regions that divide up the space (figure 7.11).

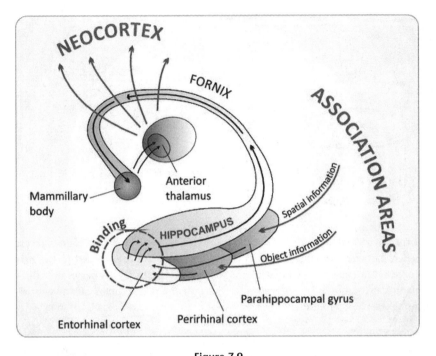

Figure 7.9

Schematic of rat hippocampus and entorhinal cortex. From https://openi.nlm.nih .gov/imgs/512/153/3370157/PMC3370157_11065_2012_9202_Fig1_HTML.png

Do the same with a bigger box and recording from a different set of neurons in the same part of the medial entorhinal cortex. They fire in the same patterns of triangles that form larger hexagons. In fact there are at least four sets of neurons that fire in hexagonal patterns for larger and larger boxes. All these neurons are located along the edge of the medial entorhinal cortex next to the hippocampus. As you record neurons' firing from the top to the bottom of the medial entorhinal cortex there is an increase in the number of neurons that fire for bigger hexagonals. But ones that record smaller size hexagons continue to be distributed among them. In this way, the experimenters were able to locate distinct clusters of neurons along the edge of the entorhinal cortex nearest the hippocampus that fired for different-sized boxes.

We need to be clear here. The dots that mark the firing of particular neurons when the rat crossed over particular spots in the boxes marked

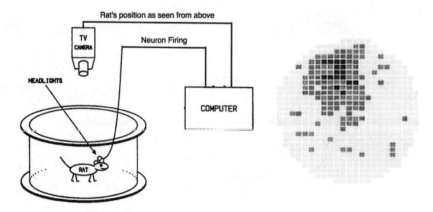

Figure 7.10

Experimental setup. As the rat moves around the cage, electrodes in its brain record when particular neurons fire; locations in the space are color coded to the neurons that fire many more times than other neurons when the rat passes over those particular locations. The experimenter can read the rat's location off the neurons firing strongest. From http://blog.brainfacts.org/wp-content/uploads/rat-in-cyl-with-map.jpg

the locations of those spots; it was these locations which formed hexagonal patterns. The neurons that fired weren't wired together into anything like hexagons. There was no one-to-one "mapping" from locations of spots in the boxes to locations of neurons in the medial entorhinal cortex of the rat's brain that would enable experimenters to "read" the shape of the space between locations of spots in the boxes from the shape of the space between the neurons that fired when the rat was at these different spots in the boxes. The experimenters found that the neurons that fired were pretty much spread out over the back of the medial entorhinal cortex, and, though called "grid neurons," these neurons *weren't in a grid themselves.*

Experimenters can draw a scaled map, an accurate representation of the rat's play space, just by watching which neurons fire in the rat's medial entorhinal cortex, to predict which neurons would fire just by watching where the rat was in its play space. It would be easy to mistake such a map of the rat's play space made by reading the recorded firing of grid cells for a map the rat or its brain might make for itself. Indeed, this is the very mistake O'Keefe made about his own earlier discovery of the place cells that connected to the grid cells. The mistake, though relatively harmless to his research progress, seriously obscures how the brain worked. In his 2014

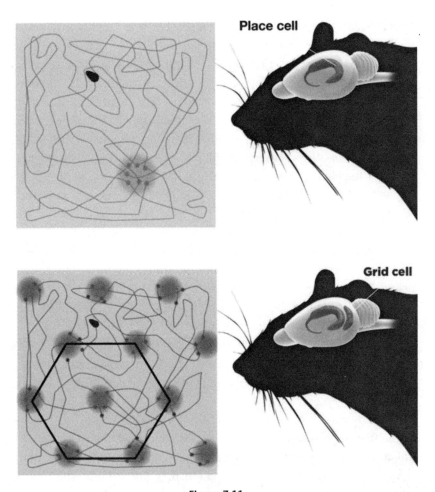

Figure 7.11

Place cells firing in the hippocampus give the rat's location in the experimental box. Grid cells firing in the medial entorhinal cortex give the geography of the box: each dot represents the firing of a single neuron. Each neuron fires preferentially when the rat is at or very near a specific location in the box. From http://youngzine.org/sites /default/files/legacy/images/mind_map.jpg

Nobel Lecture, he quoted from an earlier paper: "These findings suggest that the hippocampus provides *the rest of the brain* with a spatial reference map" (O'Keefe and Dostrovsky, 1971, p. 174, as qtd. in O'Keefe, 2014; emphasis added). O'Keefe went on to say that his discovery of place cells vindicated the theory, first proposed by behavioral psychologist Edward C. Tolman (Tolman, 1948), that "animals found their way around environments by creating internal representations, which were more complicated and interesting than the simple associations between stimuli and responses beloved of the behaviorists of the Hullian persuasion" (O'Keefe, 2014, p. 278; O'Keefe is referring to the American psychologist Clark Leonard Hull). There's a clear tipoff that what O'Keefe discovered couldn't be right. No neuroscientist has ever uncovered some other set of neurons anywhere in "the rest of" the rat brain that actually treated the firing of grid or place cells as a *representation*, a set of *symbols* arranged to correspond, by some interpretation, to reality. And no neuroscientist has even looked for the "key" or "legend" that decoded this "spatial reference map" for the rat because that's not how grid or place cells work. They aren't representations at all, at least not to the rat. As we'll see, the emerging indications of how these cells work don't vindicate O'Keefe's map metaphor. For that's what it is, a metaphor, harmless for understanding the neurology that concerned O'Keefe, but seriously misleading for everyone else.

That the place cells don't represent location for the rat or its brain is of the first importance. So far as the rat is concerned, the place cells are not *about* its location, don't *contain* any "readings" about where it is, don't *mean* "now at location *x,y* in the box." If they did, O'Keefe, the Mosers, and Kandel would have to start looking for some other part of the rat's brain that *interpreted* the place cells as being *about* location, *containing* readings, *meaning* some statement expressed in the hippocampus by a rat brain's thinking. So how does what happens in the place cells and elsewhere in the rat's brain control its behaviors?

To answer that question, we'll have to dive even more deeply into neuroscientists' findings about what is happening in the grid and place cells, what information the grid cells are sending to the place cells and the place cells to the neocortex, and how the rat brain uses that information to guide the rat's behaviors. We'll need to review some of what neuroscientists have learned about the electrochemistry of the neural circuits the Mosers discovered. That will make plain that nothing happens in the rat brain the way that the theory of mind requires.

But why should these matters be of the slightest interest to the historian or to anyone else interested in Talleyrand's biography? Well, to begin with, the implications of the neuroscientists' discoveries about the rat's brain are not limited to geographical beliefs. The place cells aren't just cells for places. So far as neuroscientists can see, these cells record a vast range of "associations": much of what the rat learns and remembers about all aspects of its environment (Manns and Eichenbaum, 2006). Second, neuroscientist have lavished all this research on rats because they recognize that most of their findings apply to humans, too. There is no evidence against and a lot of evidence for the conclusion that our much bigger brains are doing just more of the same things rats' brains are doing (Kitamura et al., 2015). Finally, what neuroscientists are learning is how our brains decide, how they choose our courses of action, in the light of our environmental circumstances and previous experiences (Yu and Frank, 2015). Surely, getting a handle on that matter should be of the greatest importance in vindicating the narrative historian's task.

In the course of fifty years at the pinnacle of French politics, Talleyrand had to make myriad critical decisions that mattered to his success, indeed his survival, and the fate of a dozen European regimes for and against whom he was working. Just to take one example, in 1807, he had to decide whether to betray his emperor, Napoléon, by entering into intrigues with the Russia tsar and the Austrian court to undermine him. Why did he make those decisions? Employing the same theory of mind, for two centuries, famous biographers have wrestled with this question, without resolving it. What was his motive? Was it venality?—Talleyrand took bribes. Calculation?—he was a survivor. Patriotism?—he claimed always to serve France first. Animus?—Napoléon had shamed him before the imperial court.

Any of these motives would make a compelling story. But do any of them trace what was actually going through Talleyrand's mind?

8 Talleyrand's Betrayal: The Inside Story

In the course of a fifty-year career across five regimes, Talleyrand changed horses in midstream many times without ever getting wet. How and why he did so is the central topic of a shelf of historical works and eight twentieth- and twenty-first-century biographies published at regular intervals since 1932. As noted in chapter 7, one of these is a classic written by another adroit political tightrope walker himself, the famous British cabinet minister Duff Cooper (Cooper, 2001 [1932]). It competes with another almost contemporaneous biography written by the great academic historian Crane Brinton (Brinton, 1963 [1936]). All the authors of these histories and biographies thought that there was a fact of the matter about what Talleyrand really had in mind with each of his machinations and that they could at least get a handle on it.

But now it can be revealed that all of their purported explanations of Talleyrand's decisions to backstab are fundamentally and profoundly wrong; they're not even close to the truth of what was going on inside his mind. That's because their explanations all relied on the theory of mind and, alas, that theory's claims are completely irrelevant to Talleyrand's actual thought processes.

How did Talleyrand's decisions emerge? To answer that question, we have to understand how any decision emerges from the mind. And to do that, we have to go back to studies of the rat's brain. But, not to worry, given how much like our brain the rat's brain is (figure 8.1, plate 3), we have little reason to doubt that we decide to act the same way the rat does.

It turns out that there is a whole lot more to place cells and grid cells than just places and grids. That's why we can be increasingly confident that the discoveries that Eric Kandel, John O'Keefe and May-Britt and Edvard

Human brain

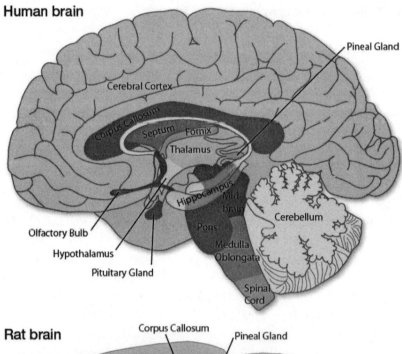

Rat brain

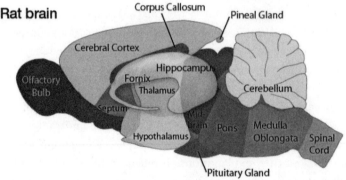

Figure 8.1
Homology of the rat and the human brain. From http://learn.genetics.utah.edu/content
/addiction/mice/brains.jpg

Mosers made generalize to tell us much more about the details of how the human as well as rat brain encodes, stores, and uses information.

Besides the grid cells, there are other cells in the medial entorhinal cortex that fire when the rat in the vicinity of barriers—walls that experimenters can put in the box to make it a rectangle or a T-maze, for example (Yu and Frank, 2015). The Mosers also identified cells that fire when a rat turns its head to its left and others that fire when the rat turns its head to its right, as well as cells that fire when the rat is walking or running at different speeds. Intermingled with the grid cells, all these other cells in the medial entorhinal cortex connect, via "projections" of synapses between neurons, to the place cells in the hippocampus (Manns and Eichenbaum, 2009) that O'Keefe discovered.

As we've seen, O'Keefe's work on place cells in the hippocampus (O'Keefe, 2014; figure 8.2, plate 4) showed how much about the rat's environment experimenters can infer from neural firing patterns. For example, there are place cells in a rat's brain that fire when the rat is in a square box and different ones that fire when it is in a circular one. Still other place cells

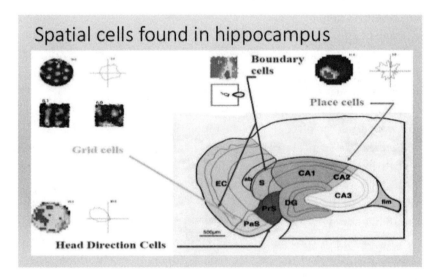

Figure 8.2
Different kinds of dedicated cells in the hippocampus of the rat brain. Grid cells, head direction cells, and boundary cells all feed into or have "projections" to the place cells; all these cell types in the rat brain hippocampus are shared by the human brain hippocampus. From O'Keefe, 2014, fig. 2.

will fire when the rat is in both a square and a circular box; of these cells, some will fire when the rat is in different locations within the box. O'Keefe suggests that place cells fire when the rat is at the same location in both a round and a square box because they are responding to the rat's location in the larger laboratory room it can see, instead of just the box in which it has been placed. Experimenters can tell in which direction on a linear track a rat is being "run" from the rate at which certain neurons fire and others don't, a pattern reversed when the rat runs in the opposite direction (Hartley et al., 2014). O'Keefe was able to establish that it takes only a small number of place cells to "cover" the entire surface of an environment, so that the hippocampus's plasticity in registering many, many different spaces is a matter of its having distinct place cell circuits that can be turned on and off by grid cell circuits (Moser et al., 2014; and by other neural circuits, as we'll see).[1]

The place cells O'Keefe discovered in the rat hippocampus get their information from the grid cells Moser found in the rat entorhinal cortex. With enough data from the grid cells, and enough computer power, experimenters can predict not only where the rat is in an experimental box, but also where it's going.

Firing of grid, head direction, speed, and barrier cells in the medial entorhinal cortex produces electrochemical disturbances along synapses connecting them to place cells. Though cells of each of these four kinds are spread all across the medial entorhinal cortex in no particular pattern, as May-Britt Moser points out in her 2014 Nobel Lecture, they all project just one synapse away from the place cells in the hippocampus (Moser, 2014).

The electrochemical firings in the entorhinal cortex dampen and strengthen one another as they produce patterns of electrochemical firing in the hippocampus's place cells. These latter cells fire in a certain pattern in response to the firings in the grid, head, speed, and barrier cells in the entorhinal cortex, as the neuroscientist says, "computing" the rat's location and direction. The neuroscientists employ the term "compute" for two reasons: first, the way the electrochemical inputs from entorhinal cortex

1. In case you're wondering what it's like to be a bat instead of a rat, in subsequent work with bats, neuroscientists discovered distinct three-dimensional place cell circuits in the bat that work the same way the two-dimensional ones do in the rat (Omer, 2018).

cells are combined in the place cells obeys a purely quantitative formula. The second reason neuroscientists describe what the hippocampal cells do as "computing" is that they are confident that they can identify the mathematics of the process and then predict the rat's movements from their, the experimenter's, representation of the states of the place cells in these mathematical equations.

In his 2014 Nobel Lecture, O'Keefe noted the way the grid cells, head direction cells, speed cells, and barrier cells of the entorhinal cortex and the place cells of the hippocampus to which they project (connect synaptically) could be used by experimenters to locate the rat the way sailors use dead reckoning to locate their ship on the sea when they can't use the stars to navigate by (O'Keefe, 2014). But "dead reckoning" is not an entirely apt metaphor since it reinforces the temptation to treat the grid, speed, head direction, and barrier cells of the medial entorhinal cortex as "representations" that the place cells "consult" to locate the rat, and, as O'Keefe found, that's not how place cells in the rat's hippocampus work. They don't treat grid cells of the entorhinal cortex as representations of anything. Nor are the place cells performing a computation on the inputs from the grid cells and then treating the result as a representation, or a symbol on a map, either. That is not what is going on.

So, what is going on? Recall that the Mosers discovered there are at least four types of grid cells, sensitive to different-sized spaces, spread across the edge of the medial entorhinal cortex. The grid cells at the dorsal (upper) end responded to short movements in small spaces. As the Mosers moved down toward the (lower) ventral end of the medial entorhinal cortex, they found that more and more distinct "modules" (the Mosers' word) of cells responded to movement across larger and larger spaces. In fact, the number of larger space modules increases over the number of nearest smaller grid modules in exactly the same ratio: 1.70 (Moser, M.-B., 2014). The four colored dots of different sizes in figure 8.3 (plate 5) remind us of this anatomical discovery. John Kubie and Steven Fox found that each set of grid cells oscillate at a distinct frequency and all synapse to the same set of place cells, where their different-frequency oscillations converge, amplify, and dampen one another into a wave peak at the place cell for a given location (Kubie and Fox, 2015).

The neurons in the rat's brain fire almost all the time. In the hippocampus, the rate of firing is between 4 and 8 cycles per second per cell; these

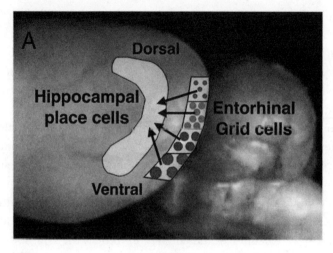

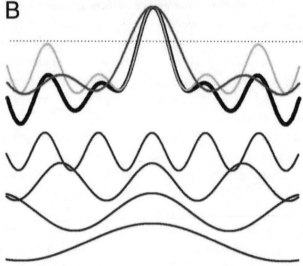

Figure 8.3
(A) Different-sized dots identify the grid cells sensitive to larger
and larger spaces in the rat's environment. (B) The waves represent
"theta wave" oscillations from the grid cells that combine at the
place cells to locate the rat. From Kubie and Fox, 2015, fig. 2.

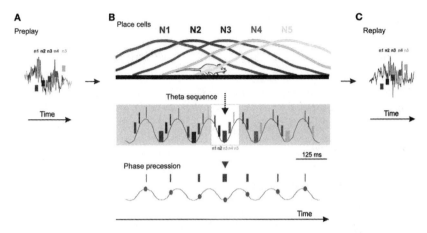

Figure 8.4

Cartoon model of theta sequences and phase precession in the CA1 area of the hippocampus. Preplay and replay panels A and C will be important later in chapter. The theta sequence and phase precession diagrams illustrate how place cells carry information about the rat's location—by firing rate of distinct cells and by when they fire in a theta oscillation. From Dragoi, 2013, fig. 1.

firings are called "theta waves." Kubie and Fox found that when the cells fire together, they produce phase oscillations in these theta waves (the blue waves in figure 8.3). Oscillations in theta waves produced by the grid and other cells in the medial entorhinal cortex combine to generate oscillations in the place cells (Kubie and Fox, 2015). O'Keefe discovered that grid cells electrochemically signal the place cells in two ways: by firing at higher rates and by the timing of individual bursts of firing against a background oscillation of circuitry firing (O'Keefe, 2014).

Figure 8.4 (plate 6) will help make this clear. In it are five place cells, N1–N5, each with a distinct "place field"—the location in the box or track where each fires. The larger circuitry in which they figure fires at a theta frequency. But each of N1–N5 fires at a slightly higher frequency and at a slightly later phase of the theta firing frequency when its particular place field in the box is crossed. The farther along in the place cell's place field, the greater the phase shift of its firing relative to the theta oscillations. The phase shift registers the rat's location.

In figure 8.4, the place cells fire in the order N1 to N5 as the rat moves down the track. The bars of varying thickness are color coded to the five cells. Their thickness denotes each cell's firing rate, and their location on the theta wave denotes the temporal sequence of their firing during the theta oscillations. The highest firing rate of each neuron is in the theta trough. Cell N3's "phase precession"—its shift in timing relative to the theta wave as the rat moves across the space where it fires—is illustrated at the bottom of the figure. That is pretty much all the signaling that's going on in the place cells—distinct electrochemical bursts in an order fixed by the rat's location as it moves on the track. (Panels A and C, marked "Preplay" and "Replay" are explained below.)

Despite the name, the place cells don't just do geography. They merge spatial and nonspatial information, "associations," like the ones Kandel got sea slugs and rats to learn and remember over time. And similarly to the grid cells, there are neural circuits elsewhere in the brain that record the nonspatial information and project it to the place cells. May-Britt Moser and others have located cells in the lateral entorhinal cortex that specialize in information about location and shape of particular objects and distinct smells (Ben-Yakov et al., 2015). These circuits work in roughly the same way that grid cells work. How else, since they are pretty well identical to grid cells in structure and wiring? The only differences between them and grid cells are to be found in what other circuits these smell and object circuits are connected to.

Clever experiments have shown that, in addition to storing information neuroscientists describe as "declarative," place cells in the rat hippocampus store information they describe as "episodic," information about an event or occasion we experienced in our past, which we can then recall or reimagine (Manns and Eichenbaum, 2009). Since rats can't tell us what episodes they remember, the evidence that the rat hippocampus records episodic information is generally based on damage to its hippocampus that prevents a rat from exploiting information about the "what, where, and when" of a previously experienced event that is normally followed by behavior the experience makes appropriate, such as successful reward seeking.

So far it has turned out that practically everything a rat needs to know about its environment works the same way: "projections" from outside the hippocampus shape the place cells, they hold the information and then transmit it through their oscillatory connections to neural circuits throughout the neocortex (Ben-Yakov et al., 2015). All the rat's "beliefs" about its environment, at least all the ones the experimentalist can reliably attribute on the basis of its behavior, start out in patterns of neural circuits firing in the rat hippocampus, patterns that then lay down the same patterns elsewhere in the neocortex (we'll see how shortly). That should come as no surprise. There's nothing special about spatial information that would make its neural coding different from other information. The only thing special about location in a box is the fact that the behavior it controls is unambiguous and easily learned. That's what made it a convenient entry point for neuroscientists to figure out how the rat's "beliefs" are formed, stored, and retrieved (as we'll see).

In all mammalian species neuroscientists have examined so far, the hippocampus appears to function in the same way—place cells, grid cells, and all the rest (Manns and Eichenbaum, 2009; Ben-Yakov et al., 2015). It's the locus where inputs from the sensory system and information stored in the neocortex come together in neural circuits that combine the information to drive appropriate behavior. The difference across species is largely a matter of which sensory modalities are dominant and so take up more space in the neocortex. For example, in the bat it's acoustical data that are preponderant inputs to the hippocampus; in us, it's visual data that are. In all these species, the hippocampus is the place where "associative" learning takes place, where stable associations between previously unrelated information from many different parts of the neocortex are established (Eichenbaum, 2013).

There is nothing like Nobel Prizes to stimulate research in a domain. Fifty years after Kandel's original work, thirty-five years after O'Keefe's, and twenty years since the Moser's, some neuroscientists are confident that "the scope of information encoded by hippocampal neurons is as broad as all attended experience" (Manns and Eichenbaum, 2006, p. 806).

Neuroscientists have found that it doesn't take many trials to train a rat to respond to a food odor emitted in one corner of a square box by going to the opposite corner for a food pellet, since after only a few trials it moves to the corner with the food immediately after smelling the odor (when it's hungry at any rate). Just as there are grid cells in the medial entorhinal cortex that store spatial information about local geography, there are cells in the lateral entorhinal cortex that store olfactory information about smells at locations; these olfactory cells feed their electrochemical signals into the place cells, just as grid cell modules do with their combined signals. Experimenters can then treat the resulting firings in the hippocampus, as *representations*.

They can predict the kind of treat present and in which corner of the rat's box it will be. And they can predict the rat's behavior based on the experimenter's representations (Moser et al., 2014; Ben-Yakov, 2015).

What's happening in the rat brain doesn't involve any representations however. All that's happening is that electrochemical oscillations are moving from the medial entorhinal cortex to the place cells (at CA1 in figure 8.1). Other oscillations are moving from the lateral entorhinal cortex to a different part of the hippocampus (CA3 in figure 8.1). Projections (synaptic connections) from the CA3 cells transmit electrochemical oscillations that combine with ones from the place cells (at CA1 in figure 8.1) to drive the rat to the food (provided there are neurons firing elsewhere in the brain as a result of low blood sugar or other causes of hunger; Eichenbaum, 2013). As we have noted, the grid cells are not a map that the place cells "consult" to figure out where the rat is. Nor do the head direction cells, or the speed cells, or the barrier cells, present the rat or some part of its brain with symbolic representations of where the rat is and where it is heading. There is no other part of the brain reading off the particular location of food in the box from the place cells that combine oscillations from the lateral and the medial entorhinal cortex. It's just one or more (usually many more) neural circuits firing one after another. To suppose otherwise is to confuse the way the rat's brain is affected by the firings of these cells with how experimenters are affected by the firing of these cells. Experimenters use recordings from the neural circuit firings as representations of, as symbols of, where the rat is and where it's heading. They use the firings to represent the information about the rat's environment, its previous training, and current location and speed, and then to predict its behavior. But neither the rat nor its brain is doing any of that. Its medial entorhinal cortex is not representing anything to its place cells. Its neurons are just causing the hippocampal ones to fire, and so on, until the rat starts moving in the direction predicted by the experimenter.

But how does the rat or, rather, its brain decide which way to move? This is our crucial question, insofar as we think Talleyrand's and our, brains work the same way rats' brains do when they decide.

The rat hippocampus is constantly forming and locally storing information about the rat's environment. But how and where does it store this information for future use? How does it retrieve it later, even much later, when it needs it to respond appropriately to environmental opportunities and threats? Remember, we must answer the same questions about how we do these things, too.

In a process called "consolidation," the hippocampus sends information for long-term storage to the neocortex in both rats and humans; it often occurs during non-REM (rapid eye movement) sleep (Roumis and Frank, 2015). The rat's hippocampus must also retrieve the stored information when it needs it, when the rat is awake, of course. Neuroscientists are beginning to understand how both tasks are accomplished (Jadhav and Frank, 2014). Recall the theta waves in the hippocampus that carry information about location and associations by rate of firing and timing of individual bursts (figures 8.3 and 8.4). In consolidation, the hippocampal neural circuits send "sharp wave ripples" (SWRs) to regions of the neocortex. It really shouldn't be a surprise that neuroscientists have found that an SWR *duplicates* the pattern of theta wave firing in the hippocampus. The surprise is that the pattern is compressed 10 times or more, from the second it took when it first occurred into milliseconds (Kay et al., 2016). The sharp wave ripple stimulates neural circuits in the much larger neocortex to take on the same pattern and theta wave firing rate that the place cell neural circuits did. The storage of association information in the neocortex frees the place cells to be reshaped by new signals from the entorhinal cortex and elsewhere. Thus a memory is shifted from the hippocampus to the neocortex by the construction of a new neural circuit with a characteristic firing rate and burst timing, one that recapitulates the pattern illustrated in figure 8.2. That's all there is to memory storage in the rat's brain—construction of an "isomorphic" (same shape—firing and bursts) neural circuit elsewhere in the brain by sharp wave ripples (Kay et al., 2016; Roumis et al., 2015).

How about the recall, retrieval, and deployment of stored information when needed? Again, the sharp wave ripples are at work here, too. Only, this time, they occur when the rat's awake and in action. These SWR bursts are called "replay" events for obvious reasons: they re-create in the hippocampus firing rates and bursts generated by previous associations recorded in the place cells. And, amazingly, these replay SWRs come in two kinds: "forward" (the pattern of the original place cell firing) and "backward" (the reverse pattern of firing; Girardeau and Zugaro, 2011). Figure 8.5 (plate 7) provides a clear illustration of several features of this neural process:

During consolidation in sleep, sharp wave ripple replay is in the forward direction only. It goes both ways during wakeful behavior. Often in the

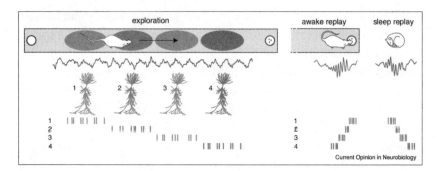

Figure 8.5

Theta waves are drawn below the track, and successive place cell circuits in the hippocampus are excited as the rat moves through different places in the track, with signature firing patters. These are recapitulated in away reverse replay starting with reward location and then recapitulated in forward replay during sleep for consolidation. From Girardeau and Zugaro, 2011, fig. 2.

forward direction when the rat is at the beginning of a training trial, and then in reverse direction, back from the location of the reward if the reward is attained. This latter process is crucial to learning by reward conditioning (Kay et al., 2016).

Both forward and backward kinds of sharp wave ripple replay are particularly prominent during learning. Indeed, SWRs are essential to learning any task that requires explicit or declarative memories, as a clever experiment by Jai Yu and Loren Frank has shown (Yu and Frank, 2015). They used a small electrical current to interfere with the production of sharp wave ripples during learning of a task that required the rat to remember (and not repeat) its previous behavior. The interference with SWR production effectively prevents the rat from learning the behavior, something that doesn't happen when experimenters interfere with the hippocampus during the rat's learning of a task requiring no declarative memory (Yu and Frank, 2015).

Once a behavioral pattern culminating in a reinforcement or a reward is learned, SWRs subside and the neocortex-hippocampus connection is maintained by much slower theta waves of the sort that feed from grid and other association cells into the hippocampus (Dragoi, 2013). Although learning and conditioning (whether classical or operant) explain much of what rats and people do, humans and especially historians don't seek explanations for learned behaviors; rather, we and they seek to explain decisions, choices, and other actions that aren't habitual, but deliberate, intentional, creative, perhaps unexpected, in need of explanation—and in which, as it turns out, SWRs play a critical role.

We can illustrate how sharp wave ripples work in decision making with a simple example. Imagine you found yourself at a "choice point," say a fork in the road in an unfamiliar location where you had to decide whether to go left or right. After looking both ways, you finally "decided" to go left. But how did that "decision" actually happen? The theory of mind says it was your beliefs and desires that selected your choice. But that's not what happened at all.

Neuroscience has figured out many of the actual details of decision making from experiments on rats. Returning to our example, while you were standing at the fork in the road trying to decide whether to go right or left, based on experimental findings in rat studies, the theta firing at the place cells in your brain would have started to produce compressed SWRs that would have replayed (in the forward direction) past firing patterns that had begun at the same or similar locations in your past experience. They would also have generated some backward replays from previously rewarded trajectories, and sometimes even random sharp wave ripples recording patterns for locations and trajectories you'd never experienced. But mainly it would have been SWRs for paths to the left and to the right that would have been generated and in roughly equal numbers. In effect, the SWRs from your hippocampus would have exited neural circuits in your neocortex to fire with rates and bursts characteristic of previous (and even possible but never tried before) movements of your body to the right and to the left (Yu and Frank, 2015).

If place cell firing during and up to your standing at that choice point produced roughly equal numbers of left and right forward SWRs, what would have made you (and experimental rats facing the same choice) eventually move in the direction replayed by one of the two trajectories and not the other? Well, it would have been just more theta waves triggered by more SWRs for one trajectory than for the other, mainly in the prefrontal cortex (PFC) and the ventral striatum, a region of the brain close to the hippocampus and directly connected to it As a result of previous experience and the resulting sharp wave ripples from the hippocampus, neural circuits would have formed in the parts of the brain that responded to the new turn left/turn right SWRs. But the responses would have been different depending on the past experiences recorded in these neural circuits of the prefrontal cortex and ventral striatum. Past experience would have resulted in some behaviors being rewarded and others not. The classically or operantly learned "associations" between choice—move left versus move right in this case—and reward would have resulted in neural circuits being preserved preferentially in these regions. In the presence of sharp wave ripples reflecting both current choices, the firing and burst patterns of previously reinforced neural circuits of the prefrontal cortex and ventral striatum would have been more strongly switched on for one choice than for the other, resulting in your (and the rats') movement left or right depending on the previous patterns of reward (Yu and Frank, 2015).

There is some evidence that the processes of this "decision making," which involves the hippocampus and the prefrontal cortex, occur in series, and not in parallel, just as our conscious experience of working memory suggests (Yu and Frank, 2015, p. 38). Each replay sharp wave ripple records firing and burst patterns from a single trajectory, so several SWRs need to be processed for the prefrontal cortex circuits to filter for the pattern among them most strongly like the previously reinforced firing/burst pattern.

If you were to face the same choice point fork in the road enough times and each time choose to go right, that choice would be rewarded until pretty soon the sharp wave ripples would disappear, and it would be just theta-wave processes that kept you going right.

There is a startling irony in these discoveries about how the brain makes choices. It entirely vindicates a research program in experimental psychology that pretty much everyone thought was consigned to oblivion fifty years ago: behaviorism. Some readers will recall the behaviorists' alternative to the theory of mind's beliefs and desires as the causes of behavior: conditioning, classical conditioning of the sort Ivan Pavlov discovered or operant conditioning ("instrumental learning") of the sort American psychologist Edward Thorndike discovered and B. F. Skinner made famous. As we noted in previous chapters, in Pavlov's classical conditioning, dogs were conditioned (learned) to salivate when a bell rang just by pairing the bell with food enough times; in operant conditioning, a behavior is conditioned (learned) when it is rewarded or punished enough times. Thorndike's "law of effect" holds that, no matter what causes a particular bit of behavior, if it is followed by a "reinforcer," it will recur with greater frequency, intensity,

or persistence, whereas if it is followed by a "punisher" or a negative reinforcer, the opposite will happen. Behaviorists thought that conditioning
was enough to explain behavior without recourse to the mental operations
or representations in the minds of animals, including humans. The influential psychologist Edward Tolman led the charge against behaviorism,
advancing the claim that behavior was the result of employing cognitive
maps (Tolman, 1948), much beloved of the Nobel laureates John O'Keefe
and May-Britt and Edvard Moser. Of course, the behaviorists were wrong,
probably more wrong than Tolman, who, at worst, mistook his map metaphor for a commitment to real representations in the rat's brain.

But the behaviorists may yet have the last laugh because it turns out that
the "whole-animal" conditioning they discovered and thought was enough
to explain human and nonhuman animal behavior really does explain it.
But it does so only when it operates on the neural circuits of the brain, all
of which appear to be built by classical or operant conditioning. Behaviorism is vindicated in the brain by a process roughly the same as the one Eric
Kandel discovered builds synapses in the hippocampus.[2]

As noted above, the story is not just one about geography. The "place"
cells in the hippocampus turn out to be "time" cells and "object location"
cells and "odor identity" cells and "noxious stimuli location" cells and cells

---

2. The behaviorists were crucially wrong about one thing: how much behavior is
"innate," not the result of experiential learning. But the error was not so much about
the behavior of the animal as a whole, but about the behavior of the neural circuitry.
Besides forward and reverse "replay" sharp wave ripples, there are others that "preplay" trajectories the rat has not ever experienced at all, firing and burst patterns for
neural circuitry that apparently reflect the "intrinsic structure" of the neural network
of the hippocampus, the results of development in region CA3. Recall figure 8.4
(plate 6), illustrating how place cell firing and burst patterns in theta waves registers
location. In that figure panel A illustrates the SWR preplay of a novel place sequence
and panel C illustrates SWR replay of the sequence in which the firing order of N1–N5
is preserved while the duration of the sequence is compressed by many times.

Remarkably, preplay place cell circuits fire more frequently than ones resulting
from familiar environments. They fire during the rat's exploration of a novel environment in an order fixed prior to its exploration, and they drive movement of the
rat when seeking a novel goal location it had not been trained to find (Azizi, Wiskott,
and Cheng, 2013). Apparently, novel topographies and tracks are encoded from pre-
established blocks of place-cell firing sequences, instead of responding to external
cues (Dragoi, 2013).

that record more abstract matters of explicit and declarative memory. And these place cells are where diverse information is put together into "associations" that drive behavior appropriate to the rat's environment. These "associations" we would describe as "beliefs" about where there is food, or noxious stimuli, or what the shortest way to a reward is. And, in all this activity, it's often the same individual place cells firing over and over, sometimes "remapping"—changing their connections to other neurons—replaying and preplaying electrochemical firing and burst patterns in compressed sequences through the same process of sharp wave ripples reaching back and forth between the hippocampus and the consolidated neural firing patterns throughout the neocortex (Roumis and Frank, 2015; Azizi et al., 2013).

It needs to be emphasized that all that's being transmitted from neural circuit to neural circuit are electrochemical bursts and their relative temporal locations in the theta waves. There's just the transmission of electrochemical bursts in sequences, *not* some signal sent by one neural circuit to be read or interpreted by another. There's nothing here for neuroscientists to decode. There's nothing here like the genetic code with its twenty "words" (the recipes for proteins and enzymes) spelled out in three of four "letters" (the nucleotides). There's nothing that reads the signals and interprets or "translates" them (the way the ribosomes do to the DNA letters). Of course, sophisticated molecular biologists will admit that their use of terms like "code" and "translate" is largely a metaphorical convenience and not to be taken literally (because the genetic code has no content; it's not *about* anything). However apt the metaphors in the case of DNA, there is much less scope for them in figuring out the electrochemical firing patterns of the neural circuits as illustrated in figures 8.2, 8.3, and 8.4. All neurons are pretty much the same in their firing. What makes a neuron a place cell in the rat brain is its location in a network between incoming stimuli and movements of the rat's body.

Well, you might try to write all this off as interesting information about rats that has little to do with us. What, after all, does the rat hippocampus have to tell us about *our* hippocampus, how it works, and what it does? As we've seen already, a very great deal, in fact, almost everything we know outside clinical medicine. Aside from the fact that the human hippocampus is bigger than the rat hippocampus, there doesn't seem to be much about our hippocampus that doesn't replicate the neural cell physiology, relevant cell types, and modes of operation of the rat's (or the monkey's, for that matter), right down to the sharp wave ripples. And if neuroscientists can't perform many of the same experiments on humans that they have performed on rats, they can at least do (and have done) some of these experiments on monkeys, whose brains are somewhere between ours and rats' brains in size, and they can also extrapolate from various mental disorders in people to hypotheses about memory and its dependence on the hippocampus. All these studies show that, in every dimension explored, the differences between rats and us seem to be a matter of degree, and that, at the level of neural circuits, there seems to be no detectable difference at all (Ben-Yakov et al., 2015; Eichenbaum, 2013). At least to a first approximation, our brains make decisions the same ways rats' brains do.

But, you may insist, we are conscious of some of our own deliberations, and so we know they are nothing like what rats go through when making choices in T-mazes. We are conscious of our beliefs, we can recall information to consciousness, we viscerally and often visually remember, sometimes in exquisite qualitative detail, episodes from our past, and bring to mind very specific facts about ourselves, others, and the world that weren't there in consciousness until we recalled them. We are conscious of our declarative beliefs regarding location and direction and of our declarative desires to be at certain places at certain times. And we can express these beliefs and desires to ourselves in subvocal sentences whose content we're aware of. So there has got to be something more going on in us than what happens in rats when they "believe" things about their environment and want to be in certain places at certain times. The nature of the rat's information storage has got to have nothing to do with the nature of human beliefs.

Alas, according to the most reliable of current neuroscientific findings, there's no reason to suppose that the "story" of how we know where we are is much different for us than it is for rats. To begin with, the individual anatomy of the neurons is the same, and the relevant anatomy of the entorhinal cortex and the hippocampus is, too. Neuroscientists know this from observation down to the molecular level and from the results of experiments and lesions, strokes, and other brain malfunctions in patients. Neurologists have narrowed down the causes of failure to have declarative knowledge of location and direction in humans to malfunction in specific parts of the brain. That means the subvocal sentences in our consciousness "about" where we are and where we're headed only report information held elsewhere. If the question is how that information is encoded, stored, and retrieved in us, the answer is pretty much the same way it is in rats—as oscillations in theta waves and sharp wave ripples.

We've already seen in chapter 6 that the play of subvocal sounds or mental images in consciousness doesn't carry along or create any intrinsic "aboutness" in the images or silent speech we're aware of. Bringing explicit and declarative memories to consciousness is just a matter of the play of subvocal silent speech and in some cases images across the "theater" of the mind. Where it comes from in the brain, what its effects in further imagery are, and its relation to subsequent behavior can't be decided by introspection.

And, for all we know, the same or at least similar things are going on in the rat's conscious experience. It's perfectly possible that even in the rat's brain, the hippocampal place cells "compute" or "path-integrate" location and direction, and compute all the rest of the associations the rat has learned, and they send them to some place in the rat's brain where the rat has conscious experience. Rats almost certainly do have experiences. But if neuroscientists can reliably and accurately predict rat behavior just from information in the grid, speed, direction, and boundary cells, then speculation about the role of conscious experience in the rat and its role adds nothing to the explanation of how the rat's brain stores and transmits information.

The fact is, even in our own case, consciousness has much less than we think to do with the way our brains acquire, store, and use information. Victims of amnesia were among the earliest sources of data about what information people can and cannot learn or remember and make use of without conscious recall. Although people with profound amnesia can recall nothing to consciousness beyond what "working memory" of the past few minutes tells them, they can nevertheless learn complex skills and deploy them long after training sessions. For example, they can learn to read mirror-image writing, or to deal with data entry, storage, and retrieval on a laptop, even to program it. They won't remember learning to do any of these things, of course. All this should not be surprising if what the amnesia victim acquires is knowledge of how to do things, skills, abilities, and dispositions, vastly more complicated versions of what the sea slug learns in conditioning experiments. But amnesic patients also demonstrate that they have "declarative knowledge"—true beliefs about past events that they tell us they don't consciously experience as beliefs before their minds (Corkin, 2013).

Recent studies, mainly employing fMRI localization of conscious and unconscious (subliminal) cognitive processing, suggests that, in humans, conscious experience is not an invariable component or even a crucial variable in acquiring or retrieving declarative memory—episodic autobiographical event memories, and semantic memory of facts (Henke, 2010). As in rats, the rapid formation and subsequent retrieval of episodic memories in humans is clearly localized to the hippocampus and neocortex and doesn't "require" consciousness. Functional magnetic resonance imaging shows that slower, learned encoding of "associations" characteristic of semantic memory results from multiple consolidation events between the hippocampus and the neocortex that move patterns of conditioning from neural circuits in the hippocampus to ones in the neocortex (Henke, 2010). In these and other cases, conscious awareness appears to function in the way suggested by the "global workspace" attention theory described in chapter 6. The silent speech sounds and imagery associations playing across consciousness are part of or effects of a process in the brain that connects distributed neural circuits in the prefrontal cortex to strengthen declarative memories (Henke, 2010). Presumably language, not consciousness, is what makes the critical difference between us and rats, in our vastly greater powers to encode, store, and retrieve declarative memories. Or maybe it only deludes us into thinking we remember so much more than rats just because they can't speak to us.

So, now we have a much better handle on the "inside story," in fact, the only even approximately correct description of Talleyrand's decision to betray Napoléon at Erfurt in 1808. Napoléon had taken him to a grand congress there to deal with Tsar Alexander. Instead, Talleyrand began to advise, then conspire with, and finally become a client of Alexander's. Why? What calculations was Talleyrand making? What was going on in that devious mind of his?

It turns out that none of the biographers got it right. None identified the pairings of beliefs and desires that moved Talleyrand into treasonable conversation with the emperor's enemies. The reason is not that there was another set of beliefs and desires Talleyrand had, which no one discerned, not even Metternich, the close student of Talleyrand's machinations. The reason is that there were no beliefs and desires anywhere inside Talleyrand's mind as he went through the process of deciding.

Alas, there was no method to his madness. No devious scheme too subtle for anyone to discern. It wasn't a matter of Talleyrand's deciding where the main chance lay and then acting upon it. The real inside story is that there was no story. Just a lot of unexciting firing of a lot of neural circuits, going back, from the moment Talleyrand broached treason with the Austrian ambassador, through decades and scores of years of classical and operant conditioning of the neural circuits in Talleyrand's brain. Standing there in whatever palace he found himself, firings in his hippocampus were sending sharp wave ripples out across his neocortex, where they stimulated one neural circuit after another, until combined with firings from the prefrontal cortex and ventral striatum, and doubtless a half dozen or more other regions of Talleyrand's brain, causing his throat, tongue, and lips to move and him to speak. No narrative to report here—just one damn electrochemical process after another.

Of course, most of us are going to be completely dissatisfied with this "inside story" of what was really going on in Talleyrand's mind. To begin with, even if the chain of events and actual shape of the neural firings, the theta waves and sharp wave ripples emanating from the hippocampus, all the way to the sounds Talleyrand produced while speaking to the ambassador of Tsar Alexander himself could be pinned down exactly, that's not what biographers or historians or most of us are interested in. We want to know the *meaning* behind those words of Talleyrand and how that meaning reflected his devious agenda. What was he thinking?

It should be clear by now that knowing everything that went on at the level of neural circuits in Talleyrand's brain won't help answer that question—there was no belief box and no desire box in action there.

The place cells don't represent current experience or anything else to (any other part of) the brain: the circuits in the neocortex don't work as beliefs about the past, whether recent or distant; the previously reinforced neural circuits in the ventral striatum don't identify future goals, end-states or rewards. That's not how any brain works.

So it would seem that the answer to our question "What exactly was Talleyrand thinking?" is simply "Get over it." The theory of mind is a snare and an illusion, seductive in its effects on our emotions. And though it's good for entertainment, it conveys no real understanding. In fact, the neural details of the brain show us why the theory of mind is such a blunt instrument: what really happens in the brain is nothing even remotely like what the theory of mind claims is going on. There's simply no way to map the brain's "boxology."

But wait, the philosopher of mind is ready to insist, surely this is a silly mistake, a sophomoric misunderstanding, a case of uninformed, crude, puerile reductionism. Of course, the brain doesn't operate in accordance with the theory of mind at its lowest level. The theory of mind describes the way the brain operates at a much higher level, well above the individual neurons or even neural circuits. Yes, yes, of course beliefs and desires are, at their lowest level, composed of neural circuits. But once brains evolved to become as big and as complicated as our brains, or even those of other higher primates, the brain processes became what the theory of mind is all about. Beliefs and desires, representations with directions of fit, content, and thoughts about things emerged and began to drive decisions, choices and actions. These beliefs and desires and all they entail are so complicated, composed of so many neural circuits, spread out across so much of the brain, that there is no chance of successfully identifying their detailed components. And no point in trying to do so either, because tracing out the neural processes that underlie human behavior won't answer the questions of historical narrative we want addressed.

In the next three chapters, we'll deal with this tempting, sophisticated, and apparently reasonable argument. We'll see how, in some respects, it's disingenuous, glossing over a huge embarrassment that contemporary philosophy faces. We'll revisit the coevolution of language with "mind reading" to diagnose the mistake the philosopher's argument makes about the brain, supposing that anywhere in it, at any level of organization, it can have thoughts *about* anything.

9 *Jeopardy!* "Question": "It Shows the Theory of Mind to Be Completely Wrong"

To which the correct contestant question would clearly seem to be "What is neuroscience?" Neuroscience teaches us that there's no room for the operation of the theory of mind at the level of the neural circuits and that understanding how the brain works at that level is enough to explain and predict the movements of the human body, or at least the rat body, with a lot more precision than using the theory of mind.

But, assuming that neuroscientists have got things right about the rat brain, how can they so confidently generalize to the human brain? Why do they assume that the vastly more complicated human brain has to work the same way the rat brain does?

They can so confidently generalize and assume because the individual neurons and neural circuits work the same way in both the human and the rat brain, because the structures of the human brain composed from them have the same relative positions and the same basic functions as (are "homologous" to) those of the rat brain, and because the only apparent differences in how the human brain and the rat brain work have to do with the far different numbers of their neurons: 86 billion versus 200 million.

But doesn't that big difference in numbers of neurons reflect a significant difference in degree? And can't a significant difference in degree mean a significant difference in kind? Many of us might strongly believe that it can. Besides, there's the argument at the end of chapter 8 that the theory of mind describes the processes going on in our brains at a far higher level than that of neural circuits. And because the theory's belief and desire boxes might well be built out of billions of neural circuits, neuroscientists will never find these boxes by looking at individual circuits or even fairly large groups of them. But there's another apparently attractive argument for saving the theory of mind from the findings of neuroscience. It starts by insisting that what a focus on individual neural circuits or even large numbers of them misses is the high level organization of diverse and distant neural circuits and modules. The ways these connected but distributed sets of neurons work to produce sensation, cognition, and emotion can't be discovered by looking at brain tissue any more than the program a computer is following can be read by looking at its motherboard.

Taking this parallel between brain and computer literally, the "high level" organization approach to how the brain works begins with the hardware versus software distinction in computer engineering. Laptops and desktops alike are perfectly physical things, composed of many parts, all the way from their hard drives down to the individual microcircuits etched into silicon chips on their motherboards. But besides their physical parts—their hardware—computers have software, which runs as programs encoded in electromagnetic charges on their hard drives: browser programs like Firefox, word processing programs like Word, spreadsheet programs like Excel, and so on. If you open up your laptop or desktop and look for the software, you won't see it. It's there all right, but not as a separate physical part. The same goes for the mind: It's software. That's why neuroscientists can't see the mind in action when they look inside the brain. But the mind's there nevertheless, running on the brain's hardware—on its neurons and neural circuits. In fact, the parallel goes much deeper. Just as the basic microcircuits of the computer send electromagnetic pulses—on/off, 0/1, low voltage/high voltage—through digital networks of semiconductors from semiconductor to semiconductor across wires, so the basic neural circuits of the brain send electrochemical pulses through digital networks of neurons, from neuron to neuron across synapses. The principal difference between computers and brains is that computers are programmed by designers and users, whereas brains are programmed first by development before and soon after birth and then by experience and education.

Thus, the mind is to the brain as computer software is to computer hardware. The mind operates in accordance with the "boxology" of the theory of mind introduced in chapter 3 (figure 9.1A) at the top; the brain runs its many programs—which all together amount to the mind—in accordance with the theory-of-mind computer flowchart (figure 9.1B) at the bottom.

A key premise of this argument is that, of the brain's many programs, one of them is described by the theory of mind. That program was downloaded

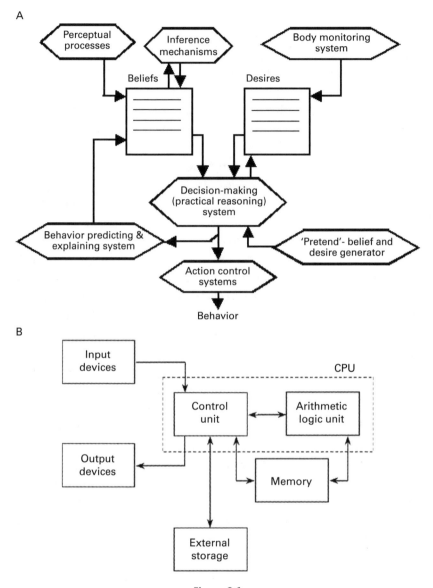

Figure 9.1

(A) "Boxology" of the theory of mind sets out marching orders for the research program of neuroscience. From Nichols et al., 1996, fig. 1. (B) Computer flowchart—computer "boxology." From http://www.superwits.com/_/rsrc/1393830517476/library /c-programming/course-content-c-programming/computeralgorithmandflowchart /ComputerBlockDiagram_Lecture1.png into all our brains at birth or soon after, like "machine code," written into the hardware of a computer at its assembly in the factory. Machine code is a computer's most basic program, needed to download, store, and execute all other programs the computer uses to process data coming in (say, from keyboard strokes or an optical scanner). The theory of mind describes a program so basic to learning (downloading) everything else that, as we saw in chapters 4 and 5, it's innate or as near to innate as Darwinian processes can contrive to make it.

Let's take this key premise of the argument seriously and see if it can help vindicate the theory of mind or at least protect it against conclusions drawn from the findings of neuroscience. The premise is roughly this: thinking is a program realized by or implemented in the neural circuits of the brain. The program proceeds roughly in accordance with the theory of mind. There are mental states—beliefs and desires—that represent with directions of fit to the world and work together to output decisions in accordance with the theory of mind's boxology. Since beliefs and desires work in accordance with a high-level program, it's impossible to detect them by examining the hardware (tissue) of the brain. With a computer, the only way to figure out what program is running is to *use* the computer: type something on its keyboard and see what comes out on the screen. Similarly, the only way to tell that the brain is running the theory-of-mind program is to stimulate the brain, watch how it interacts with its environment, or both.

Computer engineers have taken the idea of the brain as computer seriously and have turned it into at least two important breakthroughs in artificial intelligence. In exploring them, we'll see that, far from helping to reconcile the theory-of-mind claims of narrative history with the findings of neuroscience, the computer model shows what's wrong with the theory of mind in quite a different way from what neuroscience reveals.

In 1997, running a chess-playing program, an IBM mainframe computer subsequently named "Deep Blue" (figure 9.2, plate 8) was finally able to defeat the world chess champion, Gary Kasparov. The computer's program enabled it to weigh each possible move by how much it increased or decreased the probability of winning, as calculated from a large number of grand master games uploaded into its memory. But, even though its program was then fine-tuned by human grand master chess players, the computer didn't play chess the way humans do. It substituted brute force calculation plus infallible memory for human creativity. Deep Blue's chess-playing software was

Figure 9.2
Deep Blue, the IBM chess champion computer, https://i.amz
.mshcdn.com/GH65hG_jpZtWm9J_iNF9QCD0Rak=/http%
3A%2F%2Fa.amz.mshcdn.com%2Fwp-content%2Fuploads
%2F2016%2F02%2Fkasparovdeepblue-19.jpg. Courtesy of Getty.

a masterful bit of artificial intelligence, but not a computer simulation of how humans play chess.

Chess is not nearly as interesting to most people as *Jeopardy!*, the long-running American (and now international) television quiz show in which each question to contestants is posed in the form of an answer, and contestants must frame their answers in the form of questions: for example, contestants might be asked the question in the form of an answer "It's where the German army attacked its enemies four times in seventy-five years" and would have to come up with the answer in the form of a question, here, "What is the Ardennes?" *Jeopardy!* is also more interesting to computer scientists than chess is. To begin with, winning at *Jeopardy!* is harder than winning at chess: though its rules are much simpler, each player has to have a vast trove of knowledge and the ability to retrieve very specific bits by rapidly filtering through all that trove. Even the best human *Jeopardy!* players really have no idea how they search their memories, find the specifically relevant information, formulate it into questions, then speak these out loud. Thus programming a computer to play and win at *Jeopardy!* and is much harder than programming a computer to play and win at chess.

In 2011, fourteen years after Deep Blue defeated the world chess champion, an IBM computer named "Watson" (figure 9.3, plate 9; after the company's founder, Thomas J. Watson, not Sherlock Holmes's companion and scribe) defeated two human champion *Jeopardy!* players, soundly and repeatedly.

To get Watson to "understand" the *Jeopardy!* answers as questions and to produce answers to these questions in the form of questions required the computer engineers at IBM to solve a multitude of problems about how to duplicate language decoding and encoding. They had to figure out a way to feed Watson's memory circuits with an immense amount of data, including all of Wikipedia. Most important, they had to write a program that would sift through this vast database to find the right answers then speak them in the form of questions as answers to the questions posed and spoken in the form of answers. Watson is little more than electronic circuits, most of which store patterns of charges or electromagnetic fields, and some of which switch their patterns in response to electrical inputs. In that respect, Watson's electronic circuits don't much differ from the neural circuits of the sea slug or human brain. Of course, Watson runs very

Figure 9.3

Part of Watson at the IBM *Jeopardy!* studio, https://static01.nyt.com/images/2010
/06/20/magazine/20Computer-span/20Computer-span-articleLarge-v2.jpg.
Courtesy of New York Times.

complex programs that take as inputs acoustical disturbances in the form
of sound waves that are heard by humans as spoken words and that express
the quizmaster's questions (answers); it then converts these to voltage or
electromagnetic field patterns and combines them with patterns stored in
its memory to eventually produce outputs: first pushing a buzzer button
with a mechanical device and then synthesizing more acoustical disturbances to produce the spoken words of its answer in the form of a question
that the quizmaster and audience can hear. It's important to bear in mind
that everything Watson does is "algorithmic": it executes programs that are
all explicit mathematical functions stored and executed in silicon or magnetic material. Figure 9.4 shows the flowchart for the programmed steps of
how Watson responds to inputs with outputs.

So, suppose that quizmaster Alex Trebek, reading a *Jeopardy!* answer
(question) off a video screen, said, "He was born on August 15, 1769, to
an Italian family living on a French island in the Mediterranean," which
the contestants and audience would hear as they, too, read it off one of the
show's video screens. A few seconds later, and before any human contestant

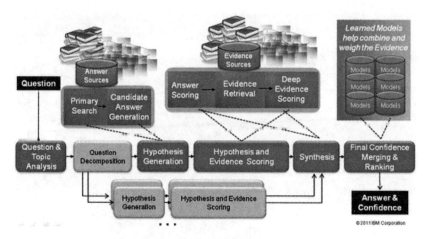

Figure 9.4

Flowchart for Watson. Everything between the question (quizmaster answer) and the answer (contestant question) is just circuits carrying charges, running programs that are purely "syntactical," and lack all content. From https://en.wikipedia.org/wiki /Watson_(computer)#/media/File:DeepQA.svg

could speak, Watson's voice synthesizer would produce the spoken words "Who was Napoléon Bonaparte?" as its question (answer).

How does Watson do it? How does it store, extract, and deploy all the information it used to defeat human contestants at *Jeopardy!*? And why suppose that any of the sequences of 0s and 1s stored in its memory that Watson used to win at *Jeopardy!* are *represented* in its electronic circuits? Well, there's really only one reason. Watson runs programs that convert the charge patterns in its memory circuits into pixels on output screens, and the computer scientists *interpret* these pixels as representing "data structures," and they write programs that convert these data structures into sounds and pixels that we non–computer scientists can understand as spoken and written words in English or whatever other language Watson is programmed to speak or write. The computer scientists have even built "peripherals"— components that can be attached to Watson enabling humans to "use" it, interact with it, interpret its outputs in spoken and written language.

That's how Watson carries the information it used to win at *Jeopardy!* Watson stores only patterns of electrical or electromagnetic charges. The computer scientists can read the information off these charges. But no matter how closely you might look at Watson's electronic circuits and the patterns of charges in its memory, you wouldn't be able to tell one data structure from another. No one could. The differences in charges don't by themselves make for any differences in the information they store. None represents Napoléon's birthday by some distinct pattern of charges "intrinsically" *about* Napoléon and the 15th of August 1769. The electronic circuits in Watson's memory differ from one another physically, just as the octagon and upside-down triangle street signs at intersections do. But Watson's circuits don't have intrinsic *content* any more than the shapes of the street signs do.

By itself, none of the many parts that make up Watson's hardware has *content, represents,* is *about* anything at all, no semiconductor, no wiring connection, no microchip, no microchip circuit. Nor does any of Watson's software either. All of it's purely formal, mathematical, "syntactical." All Watson's software programs do is change the patterns of voltages or electromagnetic fields in Watson's electronic circuits in accordance with the purely mathematical rules that the programs consist in, using lines of code that, displayed on a screen, look like those shown in figure 9.5.

As you can see in Watson's flowchart (figure 9.4), there are sixteen or so component boxes, compartments, and cylinders in the chart, each of which has a task label. But what each component box stands for is just a

Figure 9.5
Lines of code in a computer program. From http://www.mtxworld.dk/rl232-1.png

set of 1s and 0s that describes a process for changing other sets of 1s and
0s—the inputs to the program—into still other sets of 1s and 0s, which
are then sent as outputs to the next component, where more of the same
happens. The labels on the components in the flowchart identify the discrete tasks the computer scientists have broken *Jeopardy!* playing into; to
accomplish these, they've designed purely formal programs that move 0s
into 1s and vice versa.

So each of Watson's "Answer Source" or "Evidence Source" cylinders in
the figure 9.4 flowchart runs lines of code like those shown in figure 9.5.
What makes those lines of code be *about* Napoléon and his birthday? It's
the fact that computer scientists have *interpreted* them in that way.

When Watson stores information, it's just the same as when print on
the page stores information. Only it's automated. Watson doesn't intrinsically represent anywhere in its electronic circuits or in the programs it's
running that, say, Napoléon was born on August 15, 1769. Somewhere
inside Watson is a complicated distribution of electromagnetic charges
that would produce the spoken words "Who was Napoléon Bonaparte?"
the contestants and audience would hear a few seconds after Alex Trebek
would read out the *Jeopardy!* answer "He was born on August 15, 1769, to
an Italian family living on a French island in the Mediterranean," displayed
on a video screen. But Watson wouldn't know that, wouldn't believe it, any
more than the reference work from which its Napoléon data were input
by optical scanning, knows or believes anything. There's nothing inside
Watson doing any interpreting at all, nothing that's a *symbol* to Watson,
nothing that Watson would need to *interpret* as being *about* anything else.
Like the black and white marks on a printed page, the content, the meaning
of Watson's stored electronic data is derived from *our* interpretations. There
is no more intrinsic "aboutness" in Watson's microcircuits than there is in
the ink marks on this piece of paper (or the pixels on this screen if you're
reading an e-book). It's all a matter of human interpretation of the marks
and pixels.

You might think all that's obvious. Watson's programs are just more bits
of artificial intelligence, not computer simulations of what we do when we
store information in representations in our minds. But if Watson gets its
"aboutness" from us, where do *we* get *ours* from?

What's wrong with the following answer to that question? Watson's
electronic circuits are about Napoléon's birthday because a computer

scientist, call him "Bob," interprets the pixels those circuits give rise to on Watson's output screen as being *about* Napoléon. Suppose Bob was born on August 15, 1969, exactly 200 years after Napoléon's birthdate. As a result, Bob stores somewhere in his brain the fact that Napoléon and he share the same birthday. But what makes the neural circuits in Bob's brain *represent* that fact?

Well, in previous chapters, we excluded several possible answers to that question. It can't be Bob's conscious awareness of silent sounds or images that his neural circuits help produce that represent that fact. We tried out and excluded that possibility in chapter 4. And it can't be that there is some other part of Bob's brain that does what Bob does when he interprets Watson's Napoléon circuits by reading the pixels on Watson's output screen that represents that fact because it would lead to infinite regress, as we saw in chapter 7. So we excluded that possibility as well.

But here's another possible answer. Bob's neural circuits store the information that he shares a birthday with Napoléon because when he says so, his friends interpret what he says that way. Employing the theory of mind on his spoken behavior, they treat what he says as symbols for what Bob thinks, and they treat what he thinks as symbolizing in his mind the fact that he and Napoléon share a birthday. In their turn, Bob's friends' neural circuits store that same information about Bob and Napoléon because still other people employ the theory of mind on *their* behavior to interpret their thoughts, and so on *ad infinitum*.

According to this possible answer, Watson's hardware gets its content from our interpretations of what it does, and that's true of human beings, too. We are all enmeshed in a vast circle of interpretation. There is no original content anywhere to turn ink or pixels or red octagons into symbols. There are, of course, plenty of symbols, but what makes something a symbol is just our interpretation of it using other symbols.[1]

---

1. Recognizing what's going on here, some avant-garde English professors will conclude that the French deconstructionist Jacques Derrida was right when he famously said, "There's nothing outside of text [*Il n'y a pas de hors-texte*]." This statement aptly captures the thesis that there is no way to break out of the process of interpretation of one set of symbols into another. There is no original, underived, basic, brute content to the sounds or marks or pixels that we use, and none in our minds to do it either.

Derrida would have been pleased with the claim, in chapters 7 and 8, about what the Nobel Prize–winning neuroscientists discovered, where I insisted it wasn't the

The problem with this answer is not just that it begins an infinite regress, but also that, as an answer to the question "If Watson get its 'aboutness' from us, where do *we* get *ours* from?" it has little or nothing to do with the problem we're trying to solve: how the neural circuitry in our brains can act the way the theory of mind says beliefs and desires do to bring about our actions, even though it doesn't look like the circuitry works that way at all. We noticed that the on/off electronic circuits of computers work much the same way the on/off neural circuits of our brains do. So we tried out the idea that, like computers, our brains are running software. We examined how a really powerful computer like Watson does what, according to the theory of mind, our brains have to do, when we play *Jeopardy!* What we found is that, even though Watson has beaten every human contestant at playing *Jeopardy!*, there's nothing going on inside it that is anything remotely like what the theory of mind says is going on in our minds.

So, even though the idea that Watson's circuits and the programs run on these circuits represent facts because we interpret them as doing so is not wrong, it completely undermines the strategy we're trying to use to reconcile the theory of mind with the findings of neuroscience. We introduced the hardware versus software distinction to argue that the brain is hardware running software (programs) and that the theory's "boxology" describes the way one of those programs—the mind—works. Just as you can't detect the software in a computer by even the closest examination of the physical components of its hardware, so you won't see beliefs and desires, or anything that looks like them when you look at the brain's hardware, its neurons and neural circuits. But those programs are there, nevertheless, in the architecture of the brain, just like the programs in a personal computer, whether laptop or desktop, or in a mainframe computer like Watson for that matter.

---

rat that was treating the place cells and grid cells of the rat's brain as symbolic representations; it was O'Keefe and the Mosers. That the place cells and grid cells had no content for the rat, that the rat did not interpret them, is just what Derrida's thesis requires: what goes on in the rat's mind/brain doesn't break into and ground any circle of interpretation.

But, of course, no scientist or philosopher who demands evidence can take Derrida seriously. "There's nothing outside of text" was just Derrida's excuse for saying anything that came to his mind, and then rejecting obvious interpretations of what he said that made it false.

But now we've seen that programs, even programs powerful enough to win at *Jeopardy!*, don't have, indeed *can't* have, the features that the theory of mind requires. They don't operate on the content of their inputs to shape the content of their outputs. All programs do is take the form, structure, and syntax of each input and produce an output with the same or a different form, structure, and syntax. If their inputs represent anything, mean anything, are *about* anything, that fact about them doesn't matter to how programs work. Content can't matter to programs because all they can do is change 0s to 1s, low voltage to high voltage, and so on, or vice versa.

So, according to the theory of mind, what's going on in our neural circuits *can't* be a program running because the theory says it's the *content* of the representations in our brains—the beliefs and desires we actually have—that determines our decisions and drives our actions. If the computer hardware-software model is a good model of how our brains and minds work—how we think—then the theory of mind is critically, completely, and hopelessly wrong. What the theory says is going on in our brains and minds is something no program, no matter how powerful, can manage to do—operate on content, meanings, representations with directions of fit to the world. A computer program can only change 0s to 1s, and vice versa—vast quantities of them, at unimaginable speeds, with perfect accuracy. Notice, our neural circuits can do that, too; they can run "programs." It's just that those programs are nothing like what the theory of mind tells us our brains and minds are doing. Think back to what Eric Kandel, John O'Keefe, and May-Britt and Edvard Moser discovered (as we saw in chapter 7): the neural circuits of the entorhinal cortex, hippocampus, and neocortex in our brains are running a program implemented in theta waves and sharp wave ripples (SWRs), one that neuroscientists are learning more and more about. And that program has no room for interpretation, representation, content, or meaning.

That the computer model of the mind running as software on the brain's hardware isn't going to help reconcile the theory of mind with the findings of neuroscience is deeply troubling. Without such a reconciliation, narrative historical explanation, based as it is on the theory of the mind, will turn out to be little more than fantasy fiction (as we'll see in considerable detail below).

To be scientifically vindicated—the theory of mind needs a neuroscientific explanation of why it is at least roughly correct. With such an explanation, the theory would be grounded in much the same way that nineteenth-century chemical science was explained and validated by the findings of twentieth-century atomic science, or Gregor Mendel's 1866 theory of heredity was explained and validated by the findings of twentieth-century molecular biology. The two "laws" of Mendel's theory worked pretty well for his pea plants, predicting the distribution of a handful of traits from generation to generation. But there were plenty of exceptions. To be accepted as a scientifically well-grounded theory, and not just as a rule of thumb, Mendel's theory of heredity needed to establish that its proposed agents of heredity (later to be named "genes") really existed and to determine how they worked. It took almost half a century to name genes, and longer still to zero in on them—first in the cell's nucleus, then on the chromosomes in the nucleus, and then on the nucleic acid (DNA) chains that compose (part of) the chromosomes. Only then could scientists work out the structure of genes and how they operated. The great thing about James Watson and Francis Crick's identification of genes as strands of DNA in 1953 is that it didn't just vindicate the original explanations Mendel's theory provided for hereditary transmission as far back as 1866. It explained why there were exceptions to his theory and, more important, why those exceptions did not invalidate it. The findings of modern molecular biology have added greatly to the precision and the range of predictions we can make with Mendel's theory. Molecular biology has validated—and vindicated—Mendel's theory.

To see the theory of mind reconciled with, vindicated by what neuroscientists have discovered about the brain and how it works—we'd need to show (1) how the neural circuits in the brain have *content*, are *about* things, *represent* the world; and (2) how they run the rest of the body in at least roughly the way the theory of mind says beliefs and desires do.

If we can't do this, the two theories will compete with each other. The theory of mind's explanations of how the mind works will rule out neuroscience's explanations of how the brain works as being just wrong. And vice versa. We'll have to choose which theory to accept.

There's only one way the theory of mind and neuroscience could both be right, or both even at least roughly right, on track to being right about how our brains work. That way would be if neuroscientists were able to find content, "aboutness," representation somewhere in the neural circuits of the brain.

You might think there are other ways available to save the theory of mind from being proved wrong by the current findings of neuroscience. For example, you might hold that the theory of mind isn't really a theory, at least not in the formal sense, and certainly not one that has to compete with neuroscientific theory; rather, it's just a useful, informal theory, good enough for the kinds of predictions we need both to get along with other people, and to order our social and cultural institutions, our laws, even our conceptions of ourselves as free agents, where it seems both to have played an indispensable role and to have worked out well enough. You could insist that none of these purposes requires the theory of mind to be even a roughly correct scientific theory of human behavior. And if it makes claims about the causes of behavior that science can't confirm, well, that's no reason to abandon it for all those purposes. Science can't disprove or invalidate the theory of mind because it never was a scientific theory to begin with, as some philosophers have argued (Wittgenstein, 1953). In admitting that the theory of mind is not even a roughly correct scientific theory of human action, however, this position would lead us to treat it as little more than a combination of useful tool and cultural myth.

Acceptance of the cultural role of the theory of mind and of narrative history so clearly based on it without reconciling the theory with the findings of neuroscience, in effect, reduces narrative history to little more than fiction. It gives Winston Churchill's six-volume history *The Second World War* (Churchill, 1986 [1948]) roughly the status of Homer's epics, the *Iliad* and the *Odyssey,* all three of which works might well be correct recitations of events that actually happened, but whose explanations of why these events occurred do not, indeed cannot, reveal the actual causes in the brains of the people who acted in or lived through either the Second World War or the Trojan War and its aftermath. So, if narrative history's explanations are to be saved from consignment to fiction or near-fiction, neuroscientists will have to show how the neural circuits of the brain deliver what the theory of mind tells us is happening when humans take decisions and make choices.

The trouble with the project of reconciling the theory of mind with neuroscience goes back centuries. Ever since the time of René Descartes in the seventeenth century, philosophers have been wrestling with what has come to be called the "mind-body problem"—to no avail. To explain the relation between mind and brain, they have advanced a dizzying array of theories: "dualism," "occasionalism," "parallelism," "materialism," "idealism," "neutral monism," "dual aspect theory," a half dozen versions of "behaviorism," and variants on almost all of these.

By the late twentieth century, the mind-body problem had begun to absorb a great deal of philosophers' attention. It assumed its current shape just because of the advances in the new discipline of neuroscience. These advances meant that the question of reconciling the theory of mind and the findings of neuroscience could not be put off. For the last fifty years or so, the problem has taken the very form in which we face it here. Almost every philosopher of mind holds that our actions are produced by our thoughts and beliefs, and that something like the "boxology" of the theory of mind describes the way they do. Because most of these philosophers have accepted that the mind is the brain, and therefore that beliefs and desires have to be explained in terms of the workings of the brain. As a result, the problem of how to reconcile the theory of mind and the findings of neuroscience has been at the top of the research agenda of academic philosophy since neuroscience began to burgeon in the 1960s.[2]

Figuring out exactly how brain states represent with directions of fit is a problem for empirical neuroscientists, of course, and not for philosophers, who have a much harder problem to figure out: how the brain could ever possibly have content, representations, "aboutness."

The great enlightenment philosopher Leibniz made the problem clear in 1714. He dreamed up a thought experiment rather like one that Isaac Asimov, the famous science fiction writer, wrote into a novel, *Fantastic Voyage*, about 250 years later: imagine shrinking down to be so small you could enter the brain, maybe via the ear canal, and move around neural circuitry, or even move down the insides of individual nerves, or imagine you shrink to be so small you are no bigger than a neurotransmitter molecule siting between two synapses. Leibniz asked the question, What would you see? The answer is that all you'd find were individual molecules bumping into one another, changing shape, being built, and being shoved across membranes.

2. For a useful and humorous guide to the many current approaches to the reconciliation problem, see the late John Haugeland's essay "The Intentionality All-Stars," where Haugeland lays out the contrasting views by locating each at a "fielding" position in a baseball game (Haugeland, 1990). You don't need to understand baseball to grasp his points.

Of course, in 1714 Leibniz didn't know anything about the cellular anatomy of the brain, still less the electrochemistry or molecular genetics of the neurons. In his thought experiment, we are to imagine the brain increased in size so we can walk into it, as we might walk into a factory, in Leibniz' words, a "mill." All we'd find are parts pushing one another. And not even the most complicated "Rube Goldberg" contrivance of gears and cogs, springs and wheels, or for that matter genes and neurotransmitter molecules, could ever make thought (in Leibniz words) "an intelligible modification of matter," or "be comprehensible in terms of it." Leibniz went on,

> That is, a sentient or thinking being is not a mechanical thing like a watch or a mill: one cannot *conceive* of sizes and shapes and motions combining mechanically to produce something which thinks, and senses too, in a mass where [formerly] there was nothing of the kind—something which would likewise be extinguished by the machine's going out of order. *So sense and thought are not something which is natural to matter.* (Leibniz, 1981 [1765], pp. 66–67; emphasis added)

No one has ever been able to refute Leibniz's argument. And because they can't refute it, contemporary philosophers have tried to circumvent it. They've tried to show that thinking can be a brain process even though, no matter how closely we examine the brain, that process can't be detected there.

Philosophers devoted to the theory of mind have no choice but to try to circumvent Leibniz's argument, unless they're prepared to accept the "dualist" notion that mind and brain are distinct entities—that they are not one and the same thing. Hardly any philosopher nowadays wants to be a dualist. For the last fifty years or so, philosophers have twisted themselves into ever more complicated knots trying to explain how thought can possibly be physical. Today, the problem of figuring out how mental processes could ever possibly be physical is called "the hard problem" by philosophers, and the difficulty neuroscientists face in solving it is called "the explanatory gap": the inability of any neuroscientific description of what goes on in the brain to capture the experience that we have when we sense or think or feel—the experience that tells us our sensations, thoughts, and feelings have *content*, are *about* something.

Perhaps the most famous account of the "explanatory gap" was presented in 1974 by the American philosopher Thomas Nagel in his article "What Is It Like to Be a Bat?" (Nagel, 1974). Although chiropterologists have told us almost everything there is to know about how bats use echolocation to locate, track, and catch their insect prey midflight, at night, from considerable distances, and with great accuracy just by bouncing sound waves off them,[3] all the chiropterology in the world won't let us know what it's like to be a bat—won't let us think ourselves into the bat's point of view, into the *content* of its almost purely auditory consciousness of its environment. Why? Because we can't even *conceive* of what it's like to be a bat. And if we can't do that, we have nothing content-wise in the bat's experience for bat neurophysiology to explain. To see why this is so, try a version of Leibniz's thought experiment. Imagine a bat's brain could be made large enough—with all the proportions and relative places of its component regions and parts preserved—that you could actually go inside it. Once there, however, all you'd be able to know is what it's like to be *you* having your conscious (and largely visual) experiences inside the bat's brain. And you'd see almost immediately that you'd mistaken what it's like to be *you* inside the bat's brain for what it's like to *be the bat*. Nagel insisted that what it's like to be a bat is simply not expressible in any words that could describe any possible human experience. There's something about the bat's mental life that we can't explain because we can't even describe it, even though we know it's there. And, Nagel insisted, the same goes for each of us. There's something about the mental life of each of us, our conscious experience of the content and character of our sensations, thoughts, and feelings, something that words can't express, describe, capture, or convey—and something that the words of neuroscience can't explain either.

So the "hard problem" of how to bridge "the explanatory gap" really amounts to figuring out how neuroscientists could ever possibly explain how we experience our sensations, thoughts, and feelings in terms of the workings of our brains—or how they could even get started on a research program to do this. Philosophers who adopt Leibniz's stance and accept Nagel's argument insist that, no matter what neuroscientists discover about what's inside our brains, it won't answer the question of how that "what" can have the content that our consciousness tells us all our thoughts (including our unconscious ones) have.

3. In fact, the similarity of bat echolocation and human sonar is so uncanny the British chiropterologist who first uncovered bat echolocation during the Second World War was suspected of having improperly acquired defense secrets.

In other words, these philosophers consider the explanatory gap to be the hard problem that neuroscientists haven't the slightest idea how to solve. But, even though what neuroscientists like Kandel, O'Keefe, and the Mosers have found hasn't explained how the neural circuits of the rat brain have the content that the theory of mind tell us they need to have to drive rat behavior, these scientists were able to explain the complex, fine-tuned, environmentally appropriate behavior of the rat *without* attributing content, "aboutness," or meaning to the firing of neural circuits that guides that behavior. If they and others like them are able to extend their methods to the rest of the rat brain, and to the human brain, for that matter, neuroscientists will be able to explain everything the rat does, and everything we do, in ways that provide no scope for applying the theory of mind at all. Neuroscience won't need the philosophers of mind to solve the problem of how the brain can ever possibly have content. Neuroscientists will be able to get along very well by completely ignoring the theory of mind.

If the philosophers are right that beliefs and desires with content are beyond the explanatory reach of science, the project of trying to vindicate the theory of mind that narrative history is based on by grounding the theory in neuroscience won't, indeed can't, succeed. Indeed, it's conceptually impossible for us even to try to succeed.

In looking to neuroscience to vindicate the theory of mind, we're hoping that neuroscientists can (1) show why and how the brain realizes the "boxology" of the theory of mind; (2) explain why, where, and how the theory's predictions break down by identifying interfering forces; and (3) contribute to the depth and detail of the theory's explanations by giving us a more exact reading on what people believe and desire than we can get by just monitoring their behavior.

But the explanatory gap dooms all three of these hopes because the theory of mind and neuroscientific theory turn out to be logically incompatible: neuroscience requires beliefs and desires to be physical, and the theory of mind excludes that possibility. And because the two logically incompatible theories compete to explain what we do when we decide, choose, and act, they're also irreconcilable—they can't both be true at the same time.

Now, recall our original project. In chapter 6, we wanted to know what triggered the First World War. Our problem was to choose between competing pairings of beliefs and desires that purported to provide the right narrative explanation of Kaiser Wilhelm's decision to give Austria-Hungary a "blank check" to start the war.

We identified four alternative pairings of beliefs and desires, which either singly or in some combination would, if Wilhelm actually had those beliefs and desires, explain what the kaiser in fact did. All four could have been wrong, of course; there could have been some other, altogether different belief-desire pairing. But we know that, if narrative history's explanations of Wilhelm's actions stand a chance of being even approximately true, there has to be a fact of the matter about what the kaiser believed and desired. We acknowledged that the evidence that would have revealed his actual motivation—the "smoking gun"—may have been irretrievably lost. But, still, it had to have been there—those paired beliefs and desires, whatever they were, at the time the kaiser made his decision.

What could that "smoking gun" actually have been? Not what Wilhelm said or wrote certainly since the kaiser may have had excellent reason to dissimulate, and, besides, speech and writing are, like all actions, ambiguous. The only reasonable candidate for "smoking gun" here is what was actually going on in the kaiser's mind. To choose between competing belief-desire pairings, we need neuroscientists, in principle, to be able to look inside Wilhelm's brain to read the content of his actual beliefs and desires from its neural circuits. Not in practice, of course, but we need them to establish at least the possibility of doing this.

We were, in effect, hoping that neuroscience could do for the theory of mind what molecular biology did for Mendel's theory of heredity by showing that Mendel's hereditary agents exist—genes—and how they work. Watson and Crick discovered what molecules make up the genes and exactly how they carry out the functions Mendel correctly guessed they had, that made them the agents of heredity and variation. We were hoping that figuring out how the neural circuits of the brain worked would show what beliefs and desires were made of and how they operated in the brain to bring about behavior. That would have vindicated the theory of mind and thus also narrative history based on that theory the way molecular biology vindicated Mendel's theory of heredity.

But things didn't turn out that way. The theory of mind says behavior is caused by the content of what we believe and desire. Neuroscience says behavior is caused by the firing of neural circuits that *lack* content. And if the philosophers of mind are right, there's no way the neural circuits could ever give rise to the contentful states of belief and desire, representations of the way things are and might be.

It's time to revisit the "higher level of organization" argument first broached in chapter 7 and addressed at the beginning of this chapter. The argument was roughly that just because the neurology of the brain is the same at its lowest ("basement") level for sea slugs, rats, and humans, and just because the neural circuits are the same at the level of cell physiology and histology for rats and humans, it doesn't follow that there's no difference between what happens in rats' brains and what happens in ours. The vastly greater number of neurons in the human than in the rat brain must make for a significant difference in the qualitative character of cognition between rats and humans. Moreover, there must be differences between rats and humans in the way the neurons of the brain are wired together, in the higher-level architecture of the anatomically distinct parts, specialized regions, dedicated areas, and modules of the brain. Just because the cellular neurophysiology of the rat brain doesn't vindicate the theory of mind doesn't mean that the neuroanatomy of the human brain won't. The whole point of cognitive neuroscience is to uncover the higher-level mental processes of perception, emotion, and cognition, and to show how these higher-level mental processes are performed by the neural circuits of the brain. It is, according to this argument, at the higher level of these processes that the theory of mind must be tested, not at the lowest level of the neural circuits composing the brain.

Now we can see how this argument misses the point of what neuroscience reveals about the theory of mind. First of all, the Nobel Prize–winning findings reported in chapters 7 and 8 reveal that the rat brain performs cognitive processes that clearly appear to be at a pretty high level and that it does so in ways that falsify the theory of mind as an explanation of how the rat acquires, stores, and deploys beliefs. Neuroscientists may yet uncover high-level cognitive processes in rats and humans independently of determining how the neural circuits for those processes work. But what they will never uncover is how such processes, or indeed how any mental processes, can have content in the brain, can operate in any other way than as lines of code do in a computer program.

Nothing that is physical can have the features the theory of mind requires. Whether made of nucleic acids and proteins, cells, tissues, or organs, because it is a physical thing, no brain—whether of a sea slug, a rat, or a human—can have states with content, that represent, and are about things and happenings outside itself.

Instead of supporting and underwriting the theory of mind as at least on the right track, neuroscience shows that the theory is completely wrong. It shows that the most important "moving parts" in the mind that the theory describes don't exist at all.

But wait, how could the theory of mind, which works so well in everyday life, which seems indispensable, and which introspection alone shows must be operating inside our minds and brains at least, be completely wrong? The notion that abstruse, hard to understand neuroscience should force us to give it up seems preposterous. What could we possibly replace it with to order and organize our lives?

Agreed. The theory works well in everyday affairs. Its predictions about people's movements and actions in our immediate vicinity over the next few minutes and even hours are fairly good, as are some of its longer-range predictions, especially about what normal people won't do, when the theory is combined with the social, cultural, and legal institutions that have been structured over history by the employment of the selfsame theory of mind.

Giving up the theory of mind would be psychologically impossible for most people most of the time. In chapters 4 and 5, we saw exactly how and why it's drummed into us, why it's bred in our bones and as near to innate as it possibly can be. We saw that introspection seduces us into insisting that the theory operates in our own conscious thought. But in chapter 6, we also saw that introspection alone can give us no confidence that thinking has the kind of content the theory of mind requires it to have.

And, in any event, the conclusion we've been driven to doesn't really require us to give up the theory of mind most of the time. It only requires us to stop taking narrative histories seriously as conferring understanding, providing knowledge, as reliable guides to the future. The conclusion demands only that we give up believing in the explanations of narrative history.

In chapter 10, we'll go back to the evolutionary origins of the theory of mind in the coevolution of mind reading and language. We'll see how together they foisted the illusory theory of mind on us. Seeing how another completely wrong theory could maintain its grip on us for two millennia will help us see how and why the theory of mind could do so for even longer. And, once we've loosened the theory of mind's grip on us, we can explore how history is to be done without it.

10 The Future of an Illusion

In 1927, Sigmund Freud published *The Future of an Illusion*, his psychoanalytic diagnosis and history of religion, which he characterized as an illusion he attributed largely to wish fulfillment (Freud, 1989 [1927]). Acknowledging religion's universality, Freud recognized its almost unbreakable hold on us, even among atheists, the pervasive way it shapes human relations, culture, art, politics, even science.

Much of what we've now uncovered about narrative explanation suggests that the theory of mind on which it's based may also be a universal illusion, but one more basic and pervasive—and one harder to surrender. And its natural history suggests that the theory was an important source for the illusion that religion foists upon us.

Though it was the best solution that natural selection could contrive for the design problem of securing collaboration among our early ancestors on the African savanna, like most such "solutions," the theory of mind was a quick and dirty contrivance, cobbled together with what was available, by a process that couldn't wait around for better, more-adaptive components. The result was an imperfect theory, one that badly overshoots and that, once adopted, turned our ancestors and us into hyperactive agency detectors. Ever since the Pleistocene, humans have sought the causes for almost all natural processes in the desires and beliefs of an all-powerful deity to which the theory of mind could be applied, in explanation, supplication, propitiation, exculpation, and celebration. There is a well-known label for this phenomenon: "anthropomorphism." Indeed, the "overshooting" of this quick and dirty solution to the problem of getting our ancestors to collaborate, by leading people to believe in an all-powerful deity that could enforce morality among them far better than any human could, is likely to have helped the theory of mind not just to forge but also to strengthen and preserve that collaboration.

Ultimately, science would unravel at least the theory of mind's "design" argument for the existence of God, with crucial implications for the theory itself. It took 2,000 years of science, from Aristotle's time to Isaac Newton's, to rid physical nature of the all-powerful agency that the theory of mind had foisted upon it. The theory of mind was banished from the physical domain when Newton showed that motion did not reflect the ends, goals, needs, or desires of anything or anyone. After Newton, physicists recognized that there was no scope for the theory of mind in the explanation of anything that happened in the physical world. Indeed, it's reported that when Napoléon asked the great French physicist Pierre-Simon Laplace what role God had in Newtonian physics, Laplace replied, "Your majesty, I have no need of that hypothesis."

It took a further 150 years after Newton for Charles Darwin to do the same for the biological domain and another twenty years for him to finally publish this achievement, hesitating largely out of fear of its consequences for religion. Before Darwin, the only reasonable explanation for the beautiful and universal means-ends economy in the biological realm was the design of an all-powerful deity in accordance with the deity's benevolent desires and unerring beliefs.

After Darwin, hardly any scientifically educated person could take the theory of mind's explanations of adaptation seriously, which effectively spelled the end of religion's most convincing argument for the existence of God.

Notice the trajectory: first, the theory of mind was banished from the physical, then from the biological domain; now neuroscience is completing this demythologizing process for the domain of human psychology. It's casting the theory of mind out of its last redoubt in science, the human brain.

To see how natural selection has fine-tuned anatomical structure into function, but without leaving any role for purpose, let's go back to the rat's brain and to the grid cells, speed cells, direction cells, and barrier cells of its medial entorhinal cortex and the place cells of its hippocampus that we discussed in chapters 7 and 8 (figure 10.1, plate 10).

What makes the neurons in the hippocampus and the medial entorhinal cortex of the rat into *grid* cells and *place* cells—cells for location and direction? Why do they have that function, given that structurally they

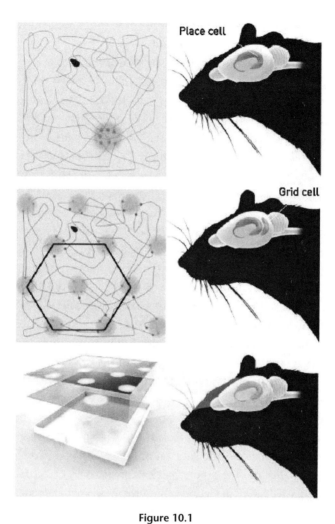

Figure 10.1

Place cells in the hippocampus and grid cells in the medial entorhinal cortex of the rat's brain. From http://www.frontline.in /multimedia/dynamic/02171/fl14_nobel_medicin_2171449g.jpg. Courtesy of Mattias Karlén/The Nobel Committee for Physiology or Medicine.

are pretty much like many other neurons throughout both the rat and the human brain?

From as early in evolution as the emergence of single-cell creatures, there was selection for any mechanism that just happened to produce environmentally appropriate behavior, such as being in the right place at the right time. In single-cell creatures, there are "organelles" that "detect" gradients in various chemicals or environmental factors (sugars, salts, heat, cold, even magnetic fields). "Detection" here simply means that, as these gradients strengthen or weaken, the organelles change shape in ways that cause their respective cells to move toward or away from the chemicals or factors as the result of some quite simple chemical reactions. Cells with organelles that happened to drive them toward sugars or away from salts survived and reproduced, carrying along these adaptive organelles. The cells whose organelles didn't respond this way didn't survive. Random variations in the organelles of other cells that just happened to convey benefits or advantages or to meet those cells' survival or reproductive needs were selected for.

The primitive organelles' detection of sugars or salts consisted in nothing more than certain protein molecules inside them changing shape or direction of motion in a chemical response to the presence of salt or sugar molecules. If enough of these protein molecules did this, the shape of the whole cell, its direction, or both would change, too. If cells contained organelles with iron atoms in them, the motion of the organelles and the cells themselves would change as soon as the cells entered a magnetic field. If this behavior enhanced the survival of the cells, the organelles responsible for the behavior would be called "magnetic field detectors." There'd be nothing particularly "detecting" about these organelles, however, or the cells they were part of. The organelles and cells would just change shape or direction in the presence of a magnetic field in accordance with the laws of physics and chemistry.

The iterative process of evolution that Darwin discovered led from those cells all the way to the ones we now identify as place and grid cells in the rat's brain. The ancestors to these cells—the earliest place and grid cells in mammals—just happened to be wired to the rest of the rat's ancestors' neurology, in ways that just happened to produce increasingly adaptive responses to the rat's ancestors' location and direction. In other mammals, these same types of cells happened to be wired to the rest of the neurology in a different way, one that moved the evolution of the animal in a less-adaptive direction. Mammals wired up in less-adaptive ways lost out in the struggle for survival. Iteration (repetition) of this process produced descendants with neurons that cause behavior that is beautifully appropriate to the rat's immediate environment. So beautifully appropriate, that causing the behavior is their function.

The function of a bit of anatomy is fixed by the particular adaptation that natural selection shaped it to deliver. The process is one in which purpose, goal, end, or aim has no role. The process is a purely "mechanical" one in which there are endlessly repeated rounds of random or blind variation followed by a passive process of environmental filtration (usually by means of competition to leave more offspring). The variation is blind to need, benefit, or advantage; it's the result of a perpetual throwing of the dice in mixing genes during sex and mutation in the genetic code that shapes the bits of anatomy. The purely causal process that produces functions reveals how Darwin's theory of natural selection banishes purpose even as it produces the appearance of purpose; the environmental appropriateness of traits with functions tempts us to confer purpose on them.

What makes a particular neuron a grid cell or a place cell? There's nothing especially "place-like" or "grid-like" about these cells. They're no different from cells elsewhere in the brain. The same goes for the neural circuits in which they are combined. What makes them grid cells and place cells are the inputs and outputs that natural selection linked them to. It is one that over millions of years wired up generations of neurons in their location in ways that resulted in ever more appropriate responses for given sensory inputs from the rat's location and direction.

Evolutionary biology identifies the function of the grid and place cells in the species *Rattus rattus* by tracing the ways in which environments shaped cells in the hippocampus and entorhinal cortex of mammalian nervous systems to respond appropriately (for the organism) to location and direction. Their having that function consists in their being shaped by a particular Darwinian evolutionary process.

But what were the "developmental" details of how these cells were wired up to do this job in each individual rat's brain? After all, rats aren't born with all their grid and place cells in place (Manns and Eichenbaum, 2006). So how do they get "tuned" up to carry continually updated environmentally appropriate information about exactly where the rat is and which way the rat needs to go for food or to avoid cats? Well, this is also a matter of variation and selection by operant conditioning in the rat brain, one in which there is no room for according these cells "purpose" (except as a figure of speech, like the words "design problem" and "selection" that are used as matters of convenience in biology even though there is no design and no active process of selection in operation at all).

Like everything else in the newborn rat's anatomy, neurons are produced in a sequence and quantity determined by the somatic genes in the rat fetus. Once they multiply, the neurons in the hippocampus and the entorhinal cortex, and many other neurons in the rat's brain as well, make and unmake synaptic connections with each other. Synaptic connections that lead to behavior rewarded by the environment, such as finding the mother's teat, are repeated and thus strengthened physically (by the process Eric Kandel discovered; Kandel, 2000). Among the connections made, many are then unmade because they lead to behaviors that are not rewarded by feedback processes that strengthen the synaptic connections physically. Some are even "punished" by processes that interrupt them. In the infant rat, the place cells make contact with the grid cells by just such a process in the first three weeks of life, enabling the rat's brain to respond so appropriately to its environment that these cells are now called "place" and "grid" cells (O'Keefe and Dostrovsky, 1979). Just as in the evolution of grid and place cells over millions of years, so also in their development in the brain of a rat pup, there is no room whatever for purpose. It's all blind variation, random chance, and the passive filtering of natural selection.

These details about how the place cells and the grid cells got their functions are important here for two reasons. First, they reflect the way that 'natural selection drives any role for a theory of mind completely out of the domain of biology, completing what Newton started for the domain of physics and chemistry. They show how the appearance of design by some all-powerful intelligence is produced *mindlessly* by purely mechanical processes (Dennett, 1995). And they make manifest that the next stage in the research program that began with Newton is the banishment of the theory of mind from its last bastion—the domain of human psychology.

Second, these details help answer a natural question to which there is a tempting but deeply mistaken answer. If the grid cells and the place cells function to locate the rat's position and direction of travel, why don't they *contain* or *represent* its location and direction? If they did, wouldn't that provide the very basis for reconciling the theory of mind with neuroscience after all? This line of reasoning is so natural that it serves in part to explain the temptation to accord content to the brain in just the way that makes the theory of mind hard to shake. By now, however, it's easy to see why this reasoning is mistaken. For one thing, if the function of the place and grid cells really makes them representations of direction and location, then every organ, tissue, and structure of an organism with a function would have the same claim on representing facts about the world.

Consider the long neck of the giraffe, whose function is to reach the tasty leaves high up in the trees that shorter herbivores can't reach, or the white coat of the polar bear whose function is to camouflage the bear from its keen-eyed seal prey in the arctic whiteness. Each has a function because both are the result of the same process of random or blind variation and natural selection that evolved the grid cells in the rat. Does the giraffe's neck being long *represent* the fact that the leaves it lets the giraffe reach are particularly tasty? Is the coat of the polar bear *about* the whiteness of its arctic environment or *about* the keen eyesight of the seals on which the bear preys? Is there something about the way the giraffe's neck is arranged that says, "There are tasty leaves high up in the trees that shorter herbivores can't reach"? Is there something about the white coat of the polar bear that *expresses* the fact that it well camouflages the bear from its natural prey, seals? Of course not.

But even though they don't *represent* anything, the long neck of the giraffe and the white coat of the polar bear are *signs*: the long neck is a sign that there are tasty leaves high in the trees on the savanna, and the white coat is a sign that the bear needs to camouflage itself from its prey in the whiteness of the arctic, the way clouds are signs that it may rain. But for the neck and coat to also be *symbols*, to represent, to have the sort of content the theory of mind requires, there'd have to be someone or something to *interpret* them as *meaning* tasty leaves or a snowy environment. Think back to why red octagon street signs are symbols of the need to stop—symbols we *interpret* as such—and not merely signs of that need.

The sign versus symbol distinction is tricky enough to have eluded most neuroscientists. The firing of a grid cell is a good sign of where the rat is. It allows the neuroscientist to make a map of the rat's space, plot where it is and where it's heading. John O'Keefe called this a "cognitive map," following Edward Tolman (1948). The "map," however, is the *neuroscientist's* representation. The rat isn't carrying a map around with it, to consult about where it is and where it's heading. Almost all neuroscientists use the word "representation," which in more general usage means "interpreted symbol," in this careless way—to describe what is actually only a reliable sign. (See Moser et al., 2014 for a nice example.) The mistake is usually harmless since neuroscientists aren't misled into searching for some other part of the brain that interprets the neural circuit firing and turns it into a representation. In fact, most neuroscientists have implicitly redefined "representation" to refer to any neural state that is systematically affected by changes in sensory input and results in environmentally appropriate output, in effect, cutting the term "representation" free from the theory of mind, roughly the way evolutionary biologists have redefined "design problem" to cut it free from the same theory.

The same process of random or blind variation and natural selection that led to the polar bear evolving its white coat is what led to us evolving our mind-reading ability. Mind reading is something the neural circuits of our brains were tuned up to do as the result of a purposeless process selecting for adaptations in an environment of predators and prey in which first vertebrates, then mammals, apes, primates, hominins, and eventually *Homo sapiens* evolved.

But how did hominins' mind-reading ability—their rather powerful, refined, and sophisticated ability to cope with predators, prey, and other hominins—give rise to the theory of mind? Together with evolutionary anthropology, paleoarchaeology, cognitive linguistics, and even philosophy of language to some extent, neuroscience has reached the point where it can give a pretty clear sketch of how the useful skill of mind reading got transformed into the theory of mind and how that theory foisted itself upon us.

Once hominins appeared and began to claw their way up the food chain on the African savanna, their ability to collaborate with one another became indispensable. This is the point at which mind reading begins to exploit another of our brains' abilities: the ability that gave rise to spoken language. What ability is that and where did it come from in evolution? We don't know the answers to either of those questions. But there are some powerful theories about what the ability consists in, due mainly to Noam Chomsky (Hauser, Chomsky, and Fitch, 2002).

Although Chomsky himself has despaired of our finding the evolution-ary origins of this ability to mind read, he thinks that another of our abil-ities is at the root of language—"recursion," the ability to build nested structures in thought (Hauser, Chomsky, and Fitch, 2002). And, indeed, recursion is ever present in language. Thus we can talk about our mothers, our mother's mothers, our mother's mother's mothers, and so on. It's an interesting and important question how many recursions the normal human brain can deal with. If you were to hear someone say, "I couldn't not help but not disagree with you more," could you tell right off whether that someone was agreeing or disagreeing with you? There are four nega-tive recursions embedded in the speaker's sentence, so what the speaker would really be saying is "I couldn't disagree with you more," right? But you had to consciously puzzle that out. With just two or three negative recursions, you could do it unconsciously. So our ability to produce and process recursive thoughts unconsciously has limits, and this ability may vary from person to person, or even from culture to culture. The Pirahã people of northwest Brazil, for example, are reported by some cultural lin-guists to have a language that lacks recursion (Everett, 2005).

But in Chomsky's view, recursion is a necessary condition for language. True languages must have grammars. Grammars are rules of recursion. Recursion had to have emerged prior to language and get harnessed together with other abilities to produce language. Chomsky holds that recursion, and thus also true language is only to be found among humans. Other spe-cies communicate, of course. But they lack recursion, so they don't have true language. Our brains acquired this ability sometime after humans split off from other primates 6 million years ago. How and when our brains' acquiring this ability was selected for no one can explain. Luckily, that's not our task here. All we need to do is sketch how our ability to mind read (an ability, not a theory) coevolved with our language ability to eventually generate the platitudes about belief, desire, decision, and action, the "folk psychology" in which the theory of mind consists.

The evolution of language among mind readers almost certainly had to have started with gestures and calls—especially warning gestures and calls that might have first emerged as purely emotional expressions of alarm. Although primatologists have helped flesh out this claim in their study of chimpanzees, we would have to assume that hominins shared alarm ges-tures and calls with the chimps and our last common ancestor. In any case, it's safe to assume that hominin protolanguage started out this way and moved from gestures and calls to grunts to something like pidgins, to full-blown languages. It's well known that mind reading isn't a prerequisite of signaling. Lots of organisms with hardly any neurology at all exchange signals. Even slime molds "decide" whether to aggregate from free-living cells into multicellular reproductive structures by exchanging information— signals—in the form of secreted macromolecules. Game theorists and philosophers of language have shown that signaling between mindless organisms can emerge by Darwinian processes alone (Skyrms, 1996, 2003). But even just a little mind reading would make the emergence of language easier and its evolution faster when cooperation was adaptive. In the long run, language and mind reading clearly coevolved, with each enhancing the power of the other through a cycle of iterated selection.

Only when language finally appeared would the mind readers be ready to turn mind reading into a theory of mind—a set of statements describing how beliefs and desires produced behavior. At that point, of course, language had already long colonized consciousness, turning introspection into a theater of silent speech and visual imagery.

There is substantial evidence that anatomically modern *Homo sapiens* emerged in Africa some 300,000 years ago (Hublin et al., 2017). Once evolutionary anthropology established this, dating a dominant mystery, the major puzzle of the disciplines of evolutionary anthropology and paleoarchaeology, arose. Why is there no sign of symbolic behavior for the next couple of hundred thousand years or more and no unambiguous evidence of language until between 50,000 and 75,000 years ago? Of course, spoken words don't fossilize so it's hard to date the emergence of language. Still, scientists are puzzled by the fact that, even though one-piece tools date back a million years or so, and complex tools are at least 75,000 years old, there's no evidence of symbolic adornment, burial markings, or anything else that bears an *interpretation,* that has *content,* that seems to be *about* something else, that *represents*, until the appearance of complex tools. This fact is part of an argument that language came late to *Homo sapiens* and emerged certainly long after our species became world-class mind readers. If this chronology is right, we can at least roughly estimate the earliest date for the advent of the theory of mind. Formulating, understanding, and using the full-blown theory of mind would require a prior grasp of the words "belief" and "desire" (in one or another human language), and since beliefs and desires are *representations*, are *about* stuff, have *content*, the theory couldn't have emerged earlier than the advent of language, say, between 50,000 and 75,000 years ago.

Of course, our ancestors were mind reading long before then, long before any of them knew anything about nouns, verbs, adjectives, adverbs, grammar, sentences, statements, and meanings in general. The ability to mind read is not a matter of deploying a consciously articulated theory of mind, expressed in language that identifies beliefs and desires and that knits them into the causes of actions by their recognized relationship as representing with directions of fit that coordinate as ends and means. Mind reading couldn't have been anything much like that, at least not at first, if more than 100,000 years was required for it to give rise to language, working with recursion, solving Darwinian survival problems on the savanna.

As we saw in chapters 4 and 5, natural selection selected for the theory of mind, laying down in our brains as innate or nearly so this quick and dirty solution to a critical evolutionary design problem, presumably in response to life or death environmental circumstances. There were probably enough obvious adaptive payoffs to small increments in acquiring the theory of mind to make its gradual emergence unsurprising.

There was at least one obvious and another less obvious design problem the solution to which would have selected for acquiring the spoken words eventually required to express the theory of mind: the need for high-quality imitation to make teaching and learning possible and the need to express and resolve conflict over behaviors with different adaptive payoffs.

In discussing two of our ancestors' profound differences from other primates in chapter 5, we saw how natural selection for the ability to imitate behaviors turned a significant maladaptation—long childhood dependency—into a significant adaptation. The need to solve the design problem of preserving survival technologies—tool making, effective edible food gathering, and prey tracking—would have selected for combinations of signaling and mind reading that would have made it possible for adults to teach and for dependent children to learn complicated tasks and processes (Sterelny, 2012). Once spoken words emerged, they could be used and combined to describe objects and actions, the next evolutionary step would be the emergence of the means to correct, improve, and communicate processes and features (Cloud, 2014). At some point, there would be selection for the words that would eventually be put together to form the theory of mind, words that would figure in sentences like "What I desire is for you to sharpen your spear this way, not that way" or "I don't believe you can catch a rabbit that way." Consider the adaptive payoffs to our ancestors in learning and teaching recursive, iterated, nested sentences that could be expressed to themselves in silent speech.

Why would the expression and resolution of conflict over behaviors with different adaptive payoffs put a premium on the emergence of words like "belief" and "desire"? Among our evolutionary forbears well back into the Pleistocene, mind reading would have been a pretty sharp tool for predicting the behavior of other animals they could see in their immediate vicinity over the immediate future, but it would have been a blunter one when they tried to use it to predict the immediate behavior of animals they couldn't see over a more distant future, and blunter still the more distant that future became. Its bluntness in these uses would have resulted sometimes from their failure to detect their predictive target's ends, toward which its behavior was directed, and sometimes from their failure to detect their target's means, the features of their immediate environment salient to the target's behavior. And sometimes, too, they'd have failed to detect both. Under these conditions, cooperating mind readers using different inputs about the means-ends behavior of their predictive targets would have had different behavioral responses. One response would usually have been more adaptive than another, and there would have been selective pressure on any mechanism that would have enabled cooperators to coordinate on that response—and thus also on the emergence of words that would have enabled cooperators to identify and resolve their disagreements. That's where a theory of mind would have been of significant help: "We both want to trap that wildebeest. You believe it's going left. But I believe it's going right. Here's why…"

Solving the critical design problem/opportunity our ancestors faced in order to survive and move up the food chain on the African savanna selected for adaptive advances in communication among hominins, advances that, by 50,000 years ago or perhaps earlier, resulted in *Homo sapiens* developing language. Sometime probably pretty late in the process, our ancestors developed the words that eventually were used to form and express a theory of mind, which then, through the workings of some Baldwin-effect mechanism, became (almost) hardwired into our brains. For maximum adaptive payoffs from the very emergence of the theory of mind, however, its expressions couldn't have operated just in spoken conversations between cooperating partners. For, even to do that, these expressions would have had to invade the mind, usually in the form of silent (subvocal) speech and even images moving through consciousness (Dennett, 1995).

As it developed over evolutionary time, spoken language gave rise to significant improvements in the abilities of our ancestors to cooperate, coordinate, and collaborate with one another in their material and social culture. One of the ways it did so was by enabling them to "coin" the terms the theory of mind now uses and eventually to frame the theory itself. And, of course, since speaking was something our ancestors had done for thousands of years, it was natural for their (and now our) general theory of why people do things, the theory of mind, to use speech to explain not only human actions but also speech itself and language in general—as we have explained meaning and reference, linguistic representation and content in all our languages.[1]

1. A fine example of how the theory of mind does this is to be found in the work of the well-known twentieth-century philosopher of language Paul Grice, who formalized the role the theory of mind is supposed to play in establishing the meaning of what people say when they speak (see, for example, Grice, 1991).

Think back to the dawn of language in the gestures and grunts of the leader of a hunt. What makes his grunt mean "Now is the time to attack." According to Grice's widely accepted model of what gives vocalizations meaning, for the leader's grunt to mean "Now is the time to attack," three iterated, nested, and recursive conditions of the grunter's state of mind had to be met: he had to want the other hunters to believe (1) that it was time to attack; (2) that he grunted because he wanted them to believe it was time to attack; and perhaps most important, (3) that it was time to attack because they believed that he grunted to get them to believe that it was time to attack. You probably have to read through these three conditions several times even to grasp them. It's also worth working out why the theory of mind holds each of these condition to be necessary for the leader's grunt to really mean "Now is the time to attack."

Imagine our lead hunter had the nervous habit of grunting involuntarily every time he thought it was the moment to attack a prey, and that he was a very successful hunter that others were prepared to follow and emulate. Well, after a while, the other hunters might come to see the leader's grunt as a sign (the way clouds are a sign that it may rain) that it was time to attack. But, in this scenario, according to the theory of mind, the leader's grunt wouldn't carry the meaning that it was time to attack any more than clouds carry the meaning that it may rain. Why wouldn't the grunt mean "Now is the time to attack" in this case? Well, let's work through the

The way the theory of mind does this job is pretty simple. It just takes the features of the words and sentences of spoken language that we seek to explain and explains them by replicating them inside our minds, in beliefs and desires, with the same labels— "content," "aboutness," "representation," "meaning," "symbol"—that we use when we speak or write about them.

Science long ago saw the problem with such explanations. They leave completely unexplained what we originally sought to explain. At some point, content, aboutness, representation, meaning, and the like have to be explained in other terms. That's where neuroscience comes in. If neuroscientists could apply the theory of mind to improve on the treatments of psychopathology, surgery, and pharmacology, as well as on our predictions of normal behavior that are based on the theory, we could all have confidence that the theory of mind was at least on the right track.

But instead of vindicating the theory of mind by unpacking beliefs and desires into their component parts in the brain's neural circuits, neuroscientists have found that the brain just doesn't work that way at all. The theory of mind is based on an illusion, one that's extremely hard to shake owing to its Darwinian pedigree. But even though it was adaptive in the past and is useful, even indispensable in everyday life, and crucial to our social institutions, culture, and the arts, the theory of mind is still fundamentally mistaken as a description of what makes us tick.

---

three conditions. The hunter didn't grunt because he wanted the others to believe that it was time to attack. His grunt was "involuntary," that is, not caused by his beliefs and desires. It's safe to assume he wanted the other hunters to believe that it was time to attack, but that's not why he grunted, so the second condition is not met. And since his grunt was not caused by his beliefs and desires, it wouldn't meet the third condition either—that he grunted because he wanted them to believe that it was time to attack because they believed that he grunted to get them to believe that it was time to attack.

Now it's pretty obvious that, at the outset of hominins' or even *Homo sapiens'* earliest attempts at some sort of vocal language, they'd need to have a full-blown theory of mind in order to accord the sounds they made the kind of meaning the theory requires. But since the theory already requires that they have a well-developed language, it's pretty clear that early humans' vocalizations couldn't meet these three conditions and so the sounds they made couldn't have the sort of meaning the theory of mind required them to have.

Can any part of the theory be reconciled with science? Can we identify the part that was adaptive for our ancestors—that helped them survive and climb from the bottom of the food chain on the African savanna to the top in a few hundred thousand years? Our ability to mind read has been around for much longer than that (Tomasello, 2014), and it still seems to work in our everyday interactions with one another. Surely, there's something going on that the theory of mind latches on to when we exercise this ability? If we can find out what that something is, perhaps we can reformulate the theory of mind, or construct a successor that will improve on it?

We've noted several times that the core application of the theory of mind, the one we use it for most and in which it works best is in coping with the behavior of other people in immediate contact with us over relatively short periods of time. The more people are hidden from our direct observation, and the longer into the future we attempt to use the theory, the more ambiguous and unreliable its predictions become. Within the domain of its optimal application, the theory works well because it treats its subjects—people and other animals, as ends-means systems and combines this assumption with information about their immediate circumstances.

The theory of mind works best when it doesn't stray very far from the mind-reading ability we share with a few other mammals. Since mind readers share their target animals' environments, they have some sensory access to what the target animals see, hear, smell, taste, and so on. Mind readers also have sensory access to their target animals' current behavior and perhaps memory access (somewhere in the hippocampus or the neocortex) to the past behavior of those and similar animals in the local environment. Mind reading, whether in predators, prey, or cooperators, is just a matter of how the brain makes a means-ends calculation from a target animal's current behavior in its environment to its behavior in the near future. Think of a lion tracking a gazelle. The lion factors the gazelle's speed and endurance and the terrain to close in and make the kill. Think of the gazelle trying to escape the lion. The gazelle factors the lion's speed and endurance and the terrain to take evasive action, which it adjusts to match the lion's pursuit. The behaviors of both lion and gazelle reflect the means-ends "calculation" that mind reading consists in.

The human mind-reading ability comes to the same thing. But we're even smarter than lions and gazelles. We have a mind-reading ability that can factor more of the shared environment into its "calculation," our memories are probably better than most other animals, at least for environmental threats and opportunities important to us, and we can pick up subtler cues than most other animals can. For example, only humans and dogs show enough white in their eyes that they can track the others' gaze and coordinate on what the others are detecting in the environment (Kaminski et al,, 2009) The importance of this small difference between dogs and us, on the one hand, and between most other animals and us, on the other, was probably a critical variation that natural selection strongly selected for to enhance cooperation and coordination first among humans, and then between dogs and humans.

Take animals with a highly developed mind-reading ability to track the means-ends behavior of other animals—conspecifics, predators, prey, and coevolved animal companions—early *Homo sapiens*, for example. Add strong selection for developing anything that will enhance cooperation among such animals, like gesturing and grunting to coordinate action. With any luck and enough time in the case of *Homo sapiens*, you'll get language, first spoken, and then even written language or at least symbolic ornamentation 125,000 years later.

Thus the development of language creates a theory of mind out of the means-ends calculation that the human brain's mind-reading ability consists in. Then that theory provides language use with a ready-made, easily understood explanation of how language comes by the meanings of its words. Meanwhile, the theory also helps *Homo sapiens* climb to the top of the African food chain and continues to work well enough in interactions both between cooperating people and between them and their predators and prey (human and nonhuman). Eventually, this adaptation overshoots and humans start to use the theory of mind beyond the immediate circumstances that selected for its adaptive improvements on their mind-reading ability. Until, finally, we begin to speculate, invent conspiracies, to narrate, tell stories and write histories.

By this point, natural selection is no longer able to do much about such overshooting and our use of mind reading outside its effective range of application. Having selected for the theory of mind's placement in the mind-reading module of our brains and the reward system for using it, it can't easily fine-tune the theory further. Nor can natural selection select for turning the theory off unless and until it becomes clearly harmful for our survival and reproduction.

So now we can see what the theory of mind gets right: it tells us that humans and the other animals it applies to are *means-ends* systems. But that's only a small part of the theory, and the rest of the theory is quite wrong. The rest of the theory tells us that humans and other animals produce their means-ends behavior via the mechanism of beliefs, which provide representation of the means, and desires, which provide representations of the ends. If this were actually the way humans and nonhuman animals produce their means-ends behavior, then by narrowing down exactly what these representations were, we'd be able to sort out good (true) explanations of what people did from bad (false) explanations. But, of course, we aren't able do this, and never have been. And now we can see why we can't.

Neuroscience shows us that there are no representations in our brains. So there's nothing there that would let us, even in principle, narrow down exactly what people believed or desired—and no way for us to filter true from false narrative explanations. Thus we end up relying on false explanations, which at best serve only to satisfy our curiosity, and the satisfaction of our curiosity becomes a criterion of explanatory success. Working backward from the means-end behavior of people, we manufacture a pairings of beliefs and desires that would be obvious in the light of the means and ends of people's actual behavior. The difference between history and biography, on the one hand, and historical fiction, on the other, is simply that in history and biography—at least if carefully done with sufficient reliable information—the actual behavior of real persons constrains the belief-desire pairings we manufacture, whereas in historical fiction, authors are permitted to take liberties with that behavior and to invent persons and events that are not real. It's no surprise that a good narrative explanation of what historical agents did satisfies our curiosity: in many cases, it's pretty much a matter of the narrative explainers, when describing what the agents actually did and how that turned out, producing pairings of beliefs and desires to explain the behaviors of those agents, which readers (or listeners) then readily accept. But those pairings, whatever they are, can't be the right ones, the ones that correctly explain the agents' behavior, because there are no right belief-desire pairings. In fact, there are no belief-desire pairings at all. There are only confabulations driven by the application of means-ends calculation—means-ends pairings—from the domain of immediate environmental and biological needs, where such calculation works well, to the domain of human affairs, where it doesn't really work at all. Which means-ends pairings are those? Well, our mind-reading ability can't discriminate very finely between means-ends pairings when they differ only slightly in their impact on immediate behavior. And the ability to mind read gets worse when it comes to predicting faraway or long-term future trajectories of means-ends behavior, which is the area where the kaiser and Talleyrand operated, the area that interests historians and biographers.

Why did the kaiser issue his "blank check" to Austria-Hungary? Well, there's no way of telling now by what ends-means processes the words that came out of Wilhelm's mouth, the documents he signed his name to, the gestures he made as his courtiers waited on his pleasure, were all held together in the kaiser's mind. The kaiser's brain, having writ, moved on quickly to some other means-ends processes. But one thing we can say with confidence is that there were no belief-desire pairings in his mind from among which anyone could have chosen the right answer to the question of why Wilhelm issued his "blank check." The whole dispute about the kaiser's actual intentions is based on a mistaken theory—the theory of mind.

So now you can see that the theory of mind is a scientific dead end and that it should go the way of other scientific dead ends. But even if you can, it'll be hard for me to convince you that a theory so useful in everyday life, so crucial to human culture and institutions, and one that's been around far longer than any other theory humans have ever contrived, roughly as long as language has been, between 50,000 and 75,000 years or so, is a dead end and that we have to let go of if we're to truly understand ourselves. But a little history of science may help.

There've been several dead ends in science: theories that were considered more or less valid and held sway for a certain period, but that were finally given up as not being even on the right track to explaining what they purported to explain. Three famous examples are alchemy, the phlogiston theory of combustion, and the Ptolemaic theory of planetary movement.

Holding sway for at least 1,500 years, alchemy explained the physical world as consisting only of the four elements of earth, air, fire, and water, and the workings of human body as actions of three bodily humors, all of which governed by the spiritual action of an all-powerful deity; it is also now perhaps best known for positing that base metals could be "transmuted" into precious metals. It would finally give way to modern atomic theory and chemistry and to the discovery and scientific description of the many natural elements. On the spectrum of these three scientific dead ends, alchemy is probably the most completely mistaken of the three, though to this day some may think it vindicated at least in part both by the success of nuclear scientists in "transmuting" certain radioactive elements into other, heavier radioactive elements through neutron bombardment and by the natural process of radioactive decay, in which radioactive elements "transmute" into other, lighter elements, both radioactive and not.

The phlogiston theory explained combustion and its results by hypothesizing the existence of a substance called "phlogiston," released during burning. On careful measurement, phlogiston turned out to have an (impossible) negative mass, couldn't be unambiguously isolated from other substances, and didn't combine with other "elements" in ways that made any sense. The phlogiston theory gave way by the end of the seventeenth century to the scientifically proven theory of combustion formulated by Antoine Lavoisier when he discovered and described the properties of oxygen. Although "phlogiston" is a now a byword for the worst explanatory hypotheses of physical science, the phlogiston theory drove the leading research program of seventeenth- and eighteenth-century chemistry, resulting in the invention of some reliable measuring devices, and some of the results of its most important experiments were preserved in the new theory of combustion, even though the phlogiston theorists deeply misunderstood their own experiments.

Dating from about 200 AD, the Ptolemaic theory of planetary motion posited an extremely complicated system to accurately account for the perceived motion of the planets and Sun around the Earth. (As we'll see, since it has few of the positive and all of the negative features of the Ptolemaic theory of planetary motion, the theory of mind needs to go the way of the Ptolemaic theory, into scientific eclipse.) The theory held that the planets traveled on quite specific paths that could be inferred from their perceived motions in the night sky. These heavenly bodies, which included the Sun, had been long distinguished from other such bodies because they were perceived to "wander" across the night sky, changing their speed and direction as they did. For this reason, they were called "planetai" by the Greeks, meaning "wanderers." For at least a thousand years, Ptolemaic theory enabled astronomers to predict with increasing accuracy the complicated paths of

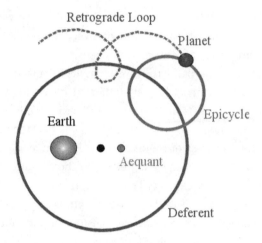

Figure 10.2

Ptolemaic model of planetary motion. From
https://en.wikipedia.org/wiki/Deferent_and
_epicycle#/media/File:Ptolemaic_elements.svg

the planets and the Sun as they moved around the Earth on "deferents" and
"epicycles."

The "deferent" was a circular orbit whose center (the black dot in figure
10.2) was near the Earth. The "epicycle" was another circular orbit whose
center moved along the deferent. The combination produced a planetary
path, showing how each planet moved across the night sky as observed
from the motionless Earth. A planet's apparent speed was originally supposed to be determined by its epicycle's distance from the "aequant" (figure
10.2), which was exactly as far from the center of the planet's orbit (the
black dot) as the Earth was. As the theory developed in the millennium
after Ptolemy, it was "improved," as a result of better data. The aequant was
replaced by more epicycles, which had "piled up" to many dozen by the
time of Nicolaus Copernicus (1473–1545). It was a pretty good theory, the
best for its time, at least in its predictive power. No surprise there. The data
to be explained were getting more refined and precise all the time and in
fact guided the "piling up" of the epicycles.

But as an explanation of the *actual* motion of the planets, the Ptolemaic theory is completely wrong. There are no epicycles, no deferents, no aequants
in the planets' actual paths. The truth—that the planets, including the
Earth, all revolve around the sun is simple ellipses—is quite different, as

Johannes Kepler, Galileo, Isaac Newton, and eventually Albert Einstein discovered. Thus a theory can be firmly believed for a very long time, can provide satisfactory predictions, indeed increasingly accurate predictions, and still be all wrong.

Like the Ptolemaic theory of planetary motion, the theory of mind finds itself having to "pile up the epicycles": adding and subtracting beliefs, revising and substituting desires, as the evidence mounts of what exactly people did and said or wrote about what they did or were going to do. The trouble for the Ptolemaic theory was that there was never an independent way of checking whether there were any epicycles and deferents at all, let alone the forty to eighty that were required to predict the paths of the planets by the time of Copernicus.[2] And the same goes for the theory of mind: since the theory was first formed thousands of years ago, there's been no way to tell whether beliefs and desires were "inscribed" on the neural circuits of the brain, and if they were, how. So there's been no independent way of checking whether any particular application of the theory of mind was right, or even on the right track. There's been no way, independent of the theory of mind, to tell what people believed and desired. The only way to infer desires and beliefs from action, speech, writing, was to use the theory of mind to reason backward from these to the belief-desire pairings that were supposed to cause them. The only way, that is, until modern neuroscience. But instead of vindicating, improving, correcting, updating, refining the theory of mind, the findings of neuroscience have shown that it's completely wrong.

But the theory of mind is actually worse off than the Ptolemaic theory, which at least could be used to reliably predict the movement of the planets and even to construct devices for calculating calendars, the timing of Olympiads and astronomical events. As I suggested above, the theory of mind is more like phlogiston theory: its set of hypotheses leads us further and further away from the right approach to explaining the phenomena in its domain. Phlogiston theory was advanced in the eighteenth century to explain the causes and effects of combustion—in particular, the observations that substances lose weight when burned. that burning quickly ceases

2. That is, there was no way until Galileo first trained his telescope on the moons of Jupiter in 1610. One look was all he needed. The moons traveled around Jupiter in simple ellipses, not in epicycles.

in enclosed spaces, that animals cannot breathe the air in enclosed spaces after burning has occurred, and that metal ores can be converted to pure metals by burning them with charcoal, which burns up almost completely. Each of these observations was explained by invoking "phlogiston," a substance that when released in burning would reduce the weight of burned substances. Burning things inside enclosed containers would fill them with phlogiston to their capacity, causing the fire to go out. Phlogiston in a closed container would kill animals placed there because it couldn't be breathed. And since it was held to be the sole ingredient of charcoal and a major ingredient of pure metals, when charcoal and metal ores were heated together, phlogiston would leave the charcoal and convert the ores into pure metals. But, with the discovery of oxygen and the role it plays in combustion and for the other reasons noted above, Lavoisier and other chemists were able to disprove all the claims of the phlogiston theory and to replace it with a correct theory of combustion.

Neuroscience has shown that, despite their appearance, human behaviors aren't really driven by purposes, ends, or goals. As in all the rest of the biological domain, there are no purposes, just a convincing illusion of purpose. Every behavior that looks like it's driven by a purpose is just the result of physical processes, like those of blind variation and natural selection uncovered by Darwin, processes that work across geological epochs to select for new species; that work in fetal development to shape tissues, organs, and innate capacities; that work to shape adaptive behavior during growth and maturation. And as neuroscience now has shown, such processes operate even on the neural circuits of our brains to drive real-time behavior.

Neuroscience is completing the scientific revolution by banishing purpose from the last domain where it's still invoked to explain and predict. Newton banished purpose from physics, and Darwin banished it from biology. Eric Kandel, John O'Keefe, May-Britt and Edvard Moser, and their fellow neuroscientists have between them begun the process of banishing purpose from psychology. John Watson, Ivan Pavlov, B. F. Skinner, and the other behaviorists tried to do this a half century or more before them. They saw that purpose has no place in the physical world. But they had no acceptable substitute for goals, ends, purposes inscribed in our brains. So, they used bluster, accusing the theory of mind of being empty, unfalsifiable, without empirical content. They were mistaken about that theory, of course. Indeed, the Ptolemaic theory of planetary of motion, the phlogiston theory of combustion, and the theory of mind all have enough empirical content to enable scientists to enlist or devise the necessary equipment to test their claims: Galileo's telescope, Lavoisier's pan balance, O'Keefe's microelectrode arrays. And because all three theories are false, science has managed to falsify them.

11 Henry Kissinger Mind Reads His Way through the Congress of Vienna

In 1956, Henry Kissinger published the results of his doctoral dissertation in an article in *World Politics*: "The Congress of Vienna: A Reappraisal" (Kissinger, 1956). Of course, Kissinger wasn't merely interested in what happened at Vienna in the months after the final defeat of Napoléon at Waterloo. If we can take his word for it, he had a much bigger target, nothing less than a reorientation of American Cold War strategy. And, of course, he succeeded in this aim: witness the consequences of this strategy over the period in which he directed and influenced American foreign policy, from 1968 until well into the 1990s.

Kissinger began that famous article by invoking the threat of nuclear extinction that faced the world in the 1950s and by condemning mere diplomacy as a vain hope of reducing tensions and averting a violent confrontation between East and West. Instead, he insisted that peace rested on much more fundamental factors:

> Nothing is more tempting than to ascribe its achievements to the very process of negotiation, to diplomatic skill, and to "willingness to come to an agreement"—and nothing is more dangerous. For the effectiveness of diplomacy depends on elements transcending it; in part on the domestic structure of the states comprising the international order, in part on their power relationship. (Kissinger, 1956, p. 264)

Warming to this theme, Kissinger lays down a series of generalizations, doctrines, and strategies that should govern a nation's foreign policy. It's worth quoting both for the breathtaking confidence, not to say grandiloquence, it reveals in a young scholar and because of the universality of his claims about peoples, nations, alliances and the international order.

> Any international settlement represents a stage in a process by which a nation reconciles its vision of itself with the vision of it by other powers. No state can

doubt its own good faith; it is the vehicle of its social cohesion. But, equally, no power can stake its survival entirely on the good faith of another; this would be an abdication of the responsibility of statesmanship. The whole domestic effort of a people exhibits an effort to transform force into obligation by means of a consensus on the nature of justice. But the international experience of a state is a challenge to the universality of its notion of justice, for the stability of the international order depends on the reconciliation of different versions of legitimacy. Could a nation achieve all its wishes, it would strive for absolute security, a world order free from the consciousness of foreign danger, and one where all problems have the manageability of domestic issues. But since absolute security for one power means absolute insecurity for all others, it is obtainable only through conquest, never as part of a legitimate settlement. An international settlement which is accepted and not imposed will therefore always appear somewhat unjust to any one of its components. Paradoxically, the generality of this dissatisfaction is a condition of stability, because were any one power totally satisfied, all others would have to be totally dissatisfied and a revolutionary situation would ensue. The foundation of a stable order is the relative security—and therefore the relative insecurity—of its members. Its stability reflects, not the absence of unsatisfied claims, but the absence of a grievance of such magnitude that redress will be sought in overturning the settlement rather than through an adjustment within its framework. An order whose structure is accepted by all major powers is "legitimate." An order containing a power which considers its structure oppressive is "revolutionary." The security of a domestic order resides in the preponderant power of authority, that of an international order in the balance of forces and in its expression, the equilibrium. (pp. 264–265)

Every sentence lays down the law. No state can do this, no power can do that, the whole domestic effort of a people always exhibits one thing, the international experience of a state is always a challenge to its notion of justice, absolute stability depends only on conquest, every peaceful settlement will appear to be unjust to someone, and so on and so forth.

There is a general theory of international relations being developed here, one that came to be called "realism" or "neorealism," derived from the German word "Realpolitik," widely used to describe nineteenth-century European diplomacy. And on what do all Kissinger's dicta depend? On history, mainly on generalization from a single historical case: the Congress of Vienna.

Kissinger thought he could convince his readers of the rightness of his approach to successful peacemaking—constructing an equilibrium between East and West—by the study of history, in particular, by examining what took place in the Congress of Vienna in 1815 and its laudable outcome:

The settlement proved all the more lasting because the negotiators at Vienna did not confuse the atmosphere of the conference table with the elements of stability of the international system. The Vienna settlement … did not rest on unsupported good faith…. Rather, there was created a structure in which the forces were sufficiently balanced…. No power felt so dissatisfied that it did not prefer to seek its remedy within the framework of the Vienna settlement rather than in overturning it…. The result was a century without a major war. (p. 280)

This is the conclusion of Kissinger's historical narrative of the Congress, a narrative he would seem to suppose establishes the rightness of his approach to peacemaking. Every step in the dance of the Vienna participants that he choreographs is driven by the theory of mind. And it's not just the beliefs and desires of men like Talleyrand and Metternich, Castlereagh and Tsar Alexander that Kissinger invokes. It's also the beliefs and desires of whole nations that figure in his narrative to move the plot along. Nor does Kissinger hesitate to put very specific thoughts in their minds:

When the British Foreign Minister, Castlereagh, spoke of the equilibrium, he meant a Europe in which hegemony was impossible; but when Metternich appealed to the equilibrium, he included a Germany in which Prussian predominance was impossible. Russia's demand for Poland threatened the equilibrium of Europe and Castlereagh could, therefore, hardly believe that any other problem was worth discussing before the Tsar's pretensions were thwarted. (p. 267)

Historians might well ask how did Kissinger know what was in Castlereagh's mind, or Metternich's—or Tsar Alexander's, for that matter? Their questions might become more insistent as Kissinger's story gathers pace, and Talleyrand takes the stage, restoring defeated France to its former status as a great power:

Prussia's insistence on Saxony merely imperiled the balance within Germany, but this was enough to absorb the full energy of Metternich. Castlereagh was interested in creating a Central Europe which would be strong enough to resist attack from both the West and the East; Metternich desired the same thing, but he was also concerned about Austria's relative position within Central Europe. To Castlereagh, the Continental nations were aspects of a defensive effort; but to the Continental nations the general equilibrium meant nothing if it destroyed the historical position which to them was the reason for their existence. To Castlereagh, the equilibrium was a mechanical expression of the balance of forces; to the Continental nations, a reconciliation of historical aspirations.

This led to a diplomatic stalemate, made all the more intractable because Britain and Austria had secured most of their special objectives during the war so that few bargaining weapons were left to Russia and Prussia, a stalemate which could be broken only by adding an additional weight to one side of the scales. Since the sole uncommitted major power was France, the former enemy emerged as the key to the European settlement. Thus grew up a myth about Talleyrand's role at the Congress of Vienna, of the diabolical wit who appeared on the scene and broke up a coalition of hostile powers, who then regrouped them into a pattern to his liking by invoking the magic word "legitimacy" and from an outcast emerged as the arbiter of Europe. To be sure, since the Treaty of Paris had settled France's boundaries, Talleyrand could afford perhaps the most disinterested approach. His wit and caustic comments became famous…. But these … would have availed little, had not the threat of France been eclipsed by the danger from the East, had not the differences among the Allies become greater than their common fear of France. So long as the Coalition still believed that the memory of the common wartime effort would provide the motive force of a settlement, Talleyrand was powerless. Once this illusion was shattered, the issue became one of the limits of self-restraint, whether a power would fail to add a factor to its side merely for the sake of the appearance of harmony. The logic of the situation provided the answer: France came to participate in European affairs, because they could not be settled without her. (pp. 267–268)

This whole story is carried forward by the theory of mind, without our even noticing it. And, of course, Kissinger the scholar would have documented—and for all we know, did document—every thought attributed to each of his players in the doctoral dissertation from which this article is drawn. But, of course, the same questions will arise about what the documents, the letters, the aide-mémoire meant, as they will arise about what the principals meant by writing them. Of every assertion Kissinger makes, we may ask the question Metternich is said to have posed about Talleyrand's death, "What did he mean by that?"

The trouble is that, if neuroscience is right, there were no beliefs and desires in any of the minds of the heads of state and their foreign ministers at Vienna. And when you "pile up the epicycles," combining more and more alleged beliefs and desires into the mind-set of whole nations, you get further and further from the truth about what actually happened. How do we combine the nonexistent beliefs and desires of individuals into the attitudes and objectives of entire nations and groupings of them in order to sensibly invoke "Prussia's insistence on gaining Saxony," "the Continental nations desire for equilibrium," "the Allies' fear of France"?

Well, isn't all this just metaphor and literary "personification"? We know perfectly well what Kissinger means here. "Russia" means the tsar; "Prussia" is Frederick William III. But Kissinger isn't writing fiction, not even historical fiction, or at least he doesn't intend to be doing so. He's committed, in effect, to there being individual belief-desire pairings behind the moves of every individual player at the Congress of Vienna, and collections, aggregations, combinations of such pairings behind every collective belief and desire, stratagem and reaction of the "powers," as they were called.

It's got to be the case that Kissinger "thinks" there are facts of the matter here since he holds that these facts about what happened at the Congress in 1815 are useful guides to policy. You can see how Kissinger's assertions about what happened at Vienna guided American policy, from invading Cambodia to end the Vietnam War, to spreading Soviet forces thinner by opening relations with the People's Republic of China, to paying the Chile's military off to eliminate its democratically elected president, to encouraging West Pakistan's murderous repression of East Pakistan.

Kissinger's historical explanation of what happened at Vienna in 1815 is fatally undermined by the absence of the very causal factors that his narrative trades in. Since none of these factors actually exists, any story that invokes them is wrong. And the same goes for Kissinger's history. Except Kissinger's story is worse since it piles up vast numbers of individual people's nonexistent belief-desire pairings into the motives, gambits and stratagems of whole nations.

So it's no surprise that his generalities and dicta about the nature of international relations and his prescriptions of how to conduct the foreign affairs of a nation went wrong as often as—or perhaps even more often than—they went right. And even when they went right, it was almost certainly an accident since they couldn't have gotten things right as a result of being driven by a theory that was not even approximately right. Combining, as Kissinger does, a large number of nonexistent pairings of beliefs and desires to explain the causes of world historical events might satisfy our curiosity and our taste for stories, and it might make what actually happened look like the inevitable result of the interplay of pairs or dozens or hundreds or millions of people with plans. But if we were to apply the "lessons" from a story such as his to cope with the future, we would almost inevitably be disappointed.

Kissinger's article illustrates how applying the theory of mind produces satisfying narratives—but, as things turned out, failed policies. According to Kissinger, in the late 1930s, the European powers facing Hitler thought that they were dealing with another Talleyrand or Metternich or Tsar Alexander, all of whom followed the prescriptions and nostrums Kissinger extracts from the success of the Congress of Vienna.

> [I]f an international order expresses the need for security and an equilibrium, it is constructed in the name of a legitimizing principle....
>
> It is the legitimizing principle which establishes the relative "justice" of competing claims and the mode of their adjustment.... Hitler's appeal to national self-determination in the Sudeten crisis in 1938 was an invocation of "justice"[:] ... it induced the Western powers to attempt to construct a "truly" legitimate order by satisfying Germany's "just" claims. Only after Hitler annexed Bohemia and Moravia [what was left of Czechoslovakia after the 1938 Munich Agreement] was it clear that he was aiming for dominion, not legitimacy; only then did the contest become one of pure power. (pp. 266–267)

It would have been a real help for Neville Chamberlain and Édouard Daladier, the British and French premiers, to know beforehand whether they were dealing with a Talleyrand or with a Hitler. The only way they could tell what Hitler was aiming for, however, would seem to have been after the fact, after Hitler had his armies swallow the rump Czechoslovakia.

It's not just that Kissinger's approach to international relations was no help in the case of Munich in 1938. Such an approach was also terribly, horribly, tragically mistaken. By acting in 1938 in the way Metternich, Castlereagh, and Talleyrand did at Vienna, where they built a century of peace, Chamberlain and Deladier got the Second World War instead. Should they have known better? Was it their historical ignorance that resulted in that tragic outcome? Quite the contrary, they would say, and Kissinger would have to agree. They knew their history all too well, having been seduced by its stories.

Where the politicians went wrong—and the historians, too—was in treating stories as knowledge and allowing them to guide policy. Neuroscientists and evolutionary psychologists have explained pretty well why we all, not just politicians and historians, treat stories as knowledge. And in the absence of any other way of explaining and predicting human actions, it's also obvious why stories would be taken as the best, or even the only, guide to policy, as Kissinger does.

But neuroscientists and their colleagues in related disciplines not only tell us why we love stories, why we remember them better than anything else, why we think or, rather, why we feel they are the easiest routes to understanding (much easier than science). Neuroscience also shows us the limits on the reliable use of stories to guide our own choices. It shows us why narrative history's stories are always wrong, and where they go wrong, and how their errors pile up in ways that make the stories it tells so often a disastrous guide to both the future and how to cope with it.

As we saw in chapter 5, natural selection contrived to make our early ancestors (and thus now us) from mind-readers into hyperactive agency detectors as a quick and dirty solution to the design problem they faced in trying to survive on the African savanna. It worked well enough right through the end of the Pleistocene and into the Holocene. But, as a predictive tool, it became blunter and blunter as human culture and its institutions developed, with its usefulness pretty well plateauing out sometime soon after agriculture and writing arose. By that time, however, it was so thoroughly bred in the bone, we were past giving it up no matter how blunt an instrument it proved to be.

Once language emerged, mind reading inevitably got mistaken for the operation of a theory of mind. Alas, that theory commits us to a quite false account of the nature and causes of human behavior, which then infects every corner of human civilization, culture, law, art, and literature. The theory is the cornerstone of the cult of reason, deliberation, distinctive human agency, free will, moral responsibility, the praiseworthiness and blameworthiness assigned by religions, law courts, and by all people who treat their fellow humans as driven by beliefs that can be evaluated for truth or falsity and by desires that can be evaluated for moral worth.

Neuroscience tells us this is all a mistake. To the extent our culture and institutions take narrative history seriously and treat its stories as knowledge, they're all built on sand, or at least on foundations that won't really support the superstructures erected on them. Modern human culture may be the best we can do, or at least the best we could have done while driven by an ability that was adapted to the environment of hunter-gatherers living in small groups with a minimal division of labor. It's easy to trace the failures of most of our political, economic, social, and cultural institutions to their dependence on a blunt instrument, mind reading, and its rationalization into a false theory, the theory of mind.

It's as though humanity never gave up the Ptolemaic theory of planetary motion, no matter how strongly the evidence piled up against it. Galileo and Newton between them not only showed that the Ptolemaic theory was wrong; they also explained why people would be tricked by their senses into believing it. They were tricked because, not feeling that the Earth was spinning, they assumed it was at rest. For them to have felt its motion, the Earth's rate of spin would have had to change. That's because feeling motion is feeling a force, and it wasn't until the seventeenth century that anyone discovered that there is no force acting on anything moving at constant velocity. So, if you're on an object traveling at constant velocity around the Earth's center of gravity, you feel that you and it are at rest. This doesn't sound like much of a discovery, but it led to the downfall of Aristotle's physics, which, for all its failures to predict much of anything, had been the only game in town for 2,000 years. Once this simple misconception became apparent, the floodgates opened to Newtonian mechanics and the scientific revolution that would banish purpose first from physics, then from biology, and perhaps soon, with neuroscience, from psychology as well. Meanwhile, mariners continued to use astronomical data and devices that Ptolemaic theory underwrote for hundreds of years more. Useful ideas take a long time to die out even when they're completely wrong. For the same reasons, no one should expect widespread surrender of the theory of mind anytime soon.

If we're to give up the theory of mind completely, we'll need at least three things. One of them we already have: an explanation of why and exactly how the theory goes wrong. A second thing we'll need is a replacement theory, one that not only avoids the theory of mind's mistakes, but that also improves on its predictions and applications in at least some important domains. A third, and perhaps most important, thing we'll need is to profoundly change our attitudes toward narrative. Even if neuroscience explains away our love of stores, we'll still pine for them, invent them, and insist on the cogency of narrative history. After all, there are people who still believe in an unmoving earth (even a flat one) and who continue to attach great weight to horoscopes cast by astrologers employing Ptolemaic theory. Some of these people have even lived in the White House in Washington, D.C.

Giving up the theory of mind is a lot harder than giving up Ptolemaic astronomy. Of the three things needed to give it up, only one is already available. And the third will probably never come to pass. Don't expect the wider culture to give up its attachment to narratives anytime soon. The entertainment value of history, historical fiction, and all of its spin-offs is far too great.

Even the explanation neuroscientists have provided us for why the theory of mind is groundless is likely to be resisted—and resisted even by neuroscientists. We can explain to ourselves why we resist the neuroscientific explanation of human behavior in the same way God-fearing evolutionary biologists have explained to themselves why they resist the theological implications of the Darwinian explanation of evolution. Many neuroscientists are reluctant to generalize from findings of other neuroscientists like Eric Kandel, John O'Keefe, May-Britt and Edvard Moser or to endorse the conclusions drawn here from their work, most notably, that there are no such things as representations with content in the brain. Sometimes such reluctance will be due to a failure to see that representation requires content. More often, it will stem from a simple implicit redefinition of "representation" that has become widespread among neuroscientists. As noted in the chapter 10, it's a redefinition that allows neuroscientists to use the word "representation" to refer to any stimulus-driven change in a neural circuit, especially when the stimulus comes from a sensory pathway.

And many neuroscientists are also reluctant to admit to the general public that neuroscience shows both the factual falsity and the practical limitations of the theory of mind. It's hard enough to get people or their representatives in government to appreciate, honor, and pay for applied research in the sciences, much less basic or fundamental research, especially when its findings and conclusions drawn from them undermine their values, challenge their cherished beliefs, and force them to rethink their worldviews. With the creationist backlash against Darwinian biology firmly in mind, perhaps the best strategy for neuroscientists, however false, wrong, or groundless they have found the theory of mind to be, is to handle that theory with kid gloves, leaving people's fundamental values, cherished beliefs, and worldviews untouched or only lightly touched, by it.

An army of philosophers of biology, biologists, and public intellectuals has sought, without notable success, to reconcile Darwinian evolutionary biology with common sense and especially with absolute moral values. A few have even sought to establish peaceful coexistence between Darwinian biology and religion, including almost all the Abrahamic and theistic denominations and groups (except perhaps for Fundamentalists who hold to the absolute inerrancy of the Bible).

Neuroscientists seeking to reconcile the theory of mind with their findings will have an easier time of it. First, it's impossible for them and almost everyone else to "switch off" their commitment to the theory of mind. So almost everyone has an incentive to reduce any cognitive dissonance produced by neuroscience in a way that saves the theory of mind. Second, for the immediate future, findings about neural circuits in the brains of laboratory animals can be passed off as irrelevant to "high-level" human cognition and conscious experience. In fact, cognitive neuroscientists have to employ the theory of mind when they give instructions to the human subjects in their experiments. Only when neuroscientific findings begin to have significant diagnostic or therapeutic payoffs in the treatment of psychological disorders beyond the reach of current magnetic imaging or stimulation technologies will neuroscientists have to face the problem of getting people to accept the implications of such findings for the theory of mind. Indeed, the impact of neuroscience's revelations about human psychology on our thinking is going to be a lot more powerful and a lot harder to deal with, but also much harder to notice—at least at first—than the impact of Darwinian revelations has been.

To at least loosen the grip of the theory of mind, to loosen the grip of narrative explanations and stories with plots, we need an approach to how we think that trades the pleasures of satisfying our curiosity for the benefits of reliability in prediction and of more effective improvement of the human condition, or at least more effective avoidance of the catastrophes that have punctuated human history until now.

The trouble with the explanation advanced in the last few paragraphs is obvious: it helps itself to the theory of mind to explain why neuroscientists are reluctant to give up that very theory. Using the theory of mind seems to be a merry-go-round we just can't get off.

But even if we can't get ourselves and everyone else to stop consuming narrative history not only for its pleasures but also for the understanding we believe it brings us, even if few people can be convinced of the wrongness of their whole conception of how all of us are driven by the conscious content of our minds, we can still hope for a theory of human psychology that enables us to better cope with the future, can't we?

12   *Guns, Germs, Steel*—and All That

In the early 1970s, an American natural scientist named Jared Diamond, who'd made his reputation studying the physiology of the gall bladder, then ecology and ornithology, came to New Guinea to study bird evolution. One day, Diamond was walking along a beach with his friend Yali, a local politician. When their conversation on how birds had colonized New Guinea shifted to how Europeans had, Yali turned to him and asked, "Why is it you white people developed so much cargo, and brought it to New Guinea, but we black people had little cargo of our own?" (qtd. in Diamond, 1997, p. 14). Taking Yali's question to heart, Diamond would put it into his own words in his 1997 book:

> Why did wealth and power become distributed as they are now, rather than in some other way? For instance, why weren't Native Americans, Africans, and Aboriginal Australians the ones who decimated, subjugated, or exterminated Europeans and Asians? (Diamond, 1997, p. 15)

Diamond was dissatisfied with the conventional answers to these questions, the ones that relied on narratives, stories about people with plots that got combined, aggregated, and built up into the fabric of academic and popular history. He made clear from the start of his book that "a history" focused on western Eurasian societies

> completely bypasses the obvious big question. Why were those societies the ones that became disproportionately powerful and innovative? The usual answers to that question invoke proximate forces, such as the rise of capitalism, mercantilism, scientific inquiry, technology, and nasty germs that kill peoples of other continents when they come into contact with western Eurasians. But why did those ingredients of conquest arise in western Eurasia, and arise elsewhere only to a lesser degree or not at all? (p. 10)

Rashly, this American ornithologist decided to try to answer that question. No "real" historian would have dreamed of undertaking such an ambitious project, having little or no use for "philosophies of history" like those of Arnold Toynbee and Oswald Spengler, much less for the efforts of amateur historians. The notion that the broad history of the world well into the twentieth century could have a single cause was simply laughable. Only a nonhistorian (who already had tenure or an independent income) would have dreamed of trying to answer Diamond's question in the compass of one book.

But when it appeared in 1997, Diamond's *Guns, Germs, and Steel* did just that. And it answered the question with such finality that there's been little or no controversy surrounding his account in the twenty or so years since its publication. It doesn't tell stories with plots; it offers no narrative linking events by the meanings agents and participants attached to them— still less by the actual beliefs and desires of those agents and participants. But his account explains not just why the Eurasians ended up dominating the globe until well into the twentieth century, but also exactly why the sequence of events played out the way it did.

Starting 13,000 years ago (some 10,000 years before the invention of writing and written history), Diamond's account of world history repeatedly invokes the same process operating on the evident local facts to drive the sequence of events he describes from the prehistory through the recorded history of humanity. The result calls to mind Darwin's *On the Origin of Species* in more ways than one. When Thomas Huxley, Darwin's "bulldog" and greatest advocate, finished reading *Origin,* he is said to have closed the book with the exclamation "How stupid of me not to have thought of that!" I had exactly the same feeling myself when I finished reading *Guns, Germs, and Steel*—as did most of its readers, judging by their reactions.

In Diamond's account, having first appeared in Africa, from about 70,000 years ago, humans then spread to the rest of the world, and their populations grew until, some 14,000 years ago, they became too large to be supported only by hunting even large wild animals and by gathering wild plants. But, after discovering how to domesticate wild plants, humans were able to cultivate them and greatly increase their food supply through farming. So agriculture arose independently in several places around the world. Of the relatively few wild plant species that could be easily domesticated, most were to be found in the Fertile Crescent—wheat, barley, protein-rich plants similar to soybeans, and flax. And, unlike the climate of most other places where agriculture took hold, the climate of the Fertile Crescent was dry for long enough each year to permit safe storage of the harvested crops. These circumstances selected for successful farming earlier in the Crescent than in other regions. Agriculture's hundredfold caloric increase in food production over hunting-gathering resulted not only in larger populations for the Fertile Crescent, but also in the spread of agriculture westward and eastward across vast regions with roughly the same climate and with fewer geographical barriers (mountain ranges, jungles, water barriers) than anywhere else in the world. Eventually, however, thousands of years of the most rapid development on Earth led to the decline both of agriculture in the Crescent and of the region itself.

The number of wild animal species that could be domesticated and exploited for food or to serve as draft or transport animals was even smaller than the number of plant species that could be domesticated and cultivated for food. But there were thirteen of these animal species in Eurasia, only one (the llama) in the Western Hemisphere, and none at all in Oceania and Australia. There may have been more such species before the rise of agriculture, but they were killed off by efficient hunter-gatherers. Zebras couldn't be domesticated nor could bison, and elephants did not readily breed in captivity. A Darwinian process of cultural selection favored human populations that were the first to succeed both in domesticating wild plants and cultivating them for food and then in domesticating wild animals and exploiting them for plowing, carting, and transport as well as for meat and milk. With both agriculture and livestock, the peoples of Eurasia would have had significant adaptive advantages over peoples everywhere else as far back as 5000 BC—and even before.

But because many of the animals they found to domesticate and exploit carried viral diseases to which humans had no immunity, domesticating them resulted in catastrophic population losses among Eurasians from smallpox, measles, and influenza. However, Eurasia was eventually repopulated by the few humans with natural immunity to these diseases and by those strong enough to survive initial infection from them. At that point domesticated animals no longer posed a threat to their survival. But, as carriers of germs to which they themselves were now immune, Eurasians and especially Europeans became a threat to the survival of peoples everywhere else.

Increasing the food supply through agriculture selected for the division of labor, which in turn led to improvements in agricultural productivity and to significant technological advances, especially in transport and weaponry. The spread of all these effects, both positive and negative, across Eurasia was assured by the long stretches of unobstructed land from the North Atlantic to the Urals and from the Urals east to the Pacific.

The European continent is characterized by alpine barriers in the interior and a crenelated coastline in the north and west. The alpine barriers to both trade and conquest provided incentives for navigation abroad, especially on the far west coast of Europe where its higher rain- and snowfall than lands to the east led to soil depletion and limitations in crop yields. All these factors selected for the projection of European power into the Atlantic, then around Africa, and finally to the Americas and Asia.

In their oceangoing ships, Europeans carried their diseases along with their draft and transport animals and weapons. Indeed, sometimes, those diseases would arrive ahead of them, sweeping in from where they'd first landed. Thus, spreading down from what would become Mexico, smallpox killed off about a third of the population of what would become Peru ten years before Francisco Pizarro got there in 1532.

Guns, Germs, and Steel illustrates how the same factors that gave rise to European agriculture, population growth, technological advances, literacy, and urbanization also occurred repeatedly and independently throughout the world. But it also identifies the factors that in each case limited such developments with the result that non-Western cultures were unable to defend themselves against European incursion, exploitation, dominance. and colonization.

It took me about 700 words to sketch out and explain the high points of Diamond's answer to the question why did wealth and power become distributed disproportionately as they are now? But *Guns, Germs, and Steel* took 450 pages of dense prose to make Diamond's case, with data, graphs, chronologies, and detailed comparisons from a dozen different disciplines. The power of his argument is repeatedly driven home by how it turns apparent exceptions into more evidence in its favor.

Having explained the broad lines of human history from the beginning of the Holocene into the twentieth century, Diamond recognizes that he has identified a program for historical research:

At present, we can put forward some partial answers plus a research agenda for the future.... The most straightforward extension of this book will be to quantify further, and thus to establish more convincingly the role of intercontinental difference in the four factors that appear to be the most important.... A second extension will be to smaller geographical scales and shorter time scales than those of this book. (pp. 408–409)

Diamond identifies his methodology as "geographical determinism," a term he appropriated from his critics and one Diamond has defended vigorously in the years since the publication of his book in 1997. "Whenever I hear the words 'geographic determinism,'" Diamond writes on his website, "I know that I am about to hear a reflex dismissal of geographic considerations, an opinion not worth listening to or reading, and an excuse for intellectual laziness and for not coming to grips with real issues" (Diamond, n.d.). He attributes historians' rejection of geographic determinism to their poor training in geography, botany, zoology, and other valuable disciplines, to the clear sense of "Western" superiority and racism that characterized earlier approaches to geopolitical analysis, and, finally, to the role of contingency and human agency that historians insist has shaped the past.

Behavioral ecologists, evolutionary anthropologists, and biologists alike will recognize that it's Darwin's theory of natural selection applied to human culture that's doing all the work in Diamond's account of history. At each stage in the process Diamond traces, it's facts about and differences between local environments that select for differences in strategies, customs, institutions. In all these different environments, humans deploy pretty much the same degree of intelligence in service of survival and reproduction. Those who happen best to exploit and take greatest advantage of the local opportunities to mitigate the local threats perpetuate their strategies while increasing their population and its dominion. As with other animal populations, the behavior of any one human population eventually changes its own environment and the environment of other populations, too. These environmental changes then select for new strategies, customs, institutions. Domestication of wild plants selects for greater productivity in supplying food and for population growth; domestication of wild animals selects for disease immunity among the domesticators, and this then selects against the populations that these germ-carrying domesticators encounter in their expansion.

The process Diamond uncovers isn't *like* Darwinian natural selection. It *is* Darwinian natural selection. In fact, Darwinian natural selection operating on the growth, succession, and extinction of cultures, civilizations, and peoples is the only possible mechanism that could result in the process *Guns, Germs, and Steel* reveals. Diamond recognizes that, in addition to the environmental conditions—the distribution of wild plants and animals that could be readily domesticated, diseases, the long, unbroken east–west stretches of land, and the barriers to north–south movement, he needs something else, the very thing that the Darwinian process needs: exogenous environmental events—contingencies or shocks—that shift the direction of natural selection into radically new paths.

In the biological realm, such events are sometimes obvious in visible ways: continental drift, earthquakes, volcanic eruptions, cometary impact. Sometimes they're invisible, at least to the naked eye: point mutations, gene duplications, chromosomal recombinations, the chance survival of one offspring instead of another, the mating of two organisms with complementary traits, the development of a symbiotic or mutualist relationship.

To see the same process at work in human history and its role in Diamond's account, consider his treatment of the eclipse of China by the West. Before the expansion of the West began, China had many advantages: efficient food production and animal domestication in an extensive temperate empire with a large population and, most of all, a huge technological advantage over the West. "The long list of its major technological firsts includes cast iron, the compass, gunpowder, paper, printing…. It also led the world in political power, navigation, control of the seas…. Why, in brief, did China lose its technological lead to the formerly so backward Europe?" (p. 412). Diamond locates the cause in an unpredictable, exogenous shock to the process traced in his book. The eclipse of China as a competitor that might have challenged the West was "a result of a typical aberration of local politics that could happen anywhere in the world: a power struggle between two factions at the Chinese court (the eunuchs and their opponents)" was won by those who put a complete stop to its already successful open ocean expansion in the early fifteenth century. As an advanced politically centralized government, the winning faction imposed its writ across the entire nation and every seaport. By contrast, Diamond identifies at least five different powers in Europe that Christopher Columbus approached to finance his voyage, until one among these fragmented, disorganized, isolated, warring states finally agreed. "Precisely because Europe was fragmented, Columbus succeeded on his fifth try in persuading one of Europe's hundreds of princes to sponsor him" (p. 412).

Just as in biological evolution, the process is the result of the repeated occurrence of "exogenous" environmental events—the arrival of bubonic plague in Europe, or point mutations—the political struggle in the Chinese court or Columbus's convincing King Ferdinand and Queen Isabella of Castile and Aragon instead of the Count of Anjou to finance his expedition. These are events, large and small, that Diamond's theory treats as driving factors but doesn't explain, just as Darwin's theory is silent on the geology and genetics that drive evolution.

Once we see what the real machinery is behind the historical process Diamond so brilliantly details, we can be confident about its application "to smaller geographic scales and shorter time periods," as Diamond suggests. Variation and selection operate on human populations and the environments they live in, make and remake them, to shape their history over 13,000 or 1,300 years, over 130 or 13 years—or 13 months, for that matter. It's just harder to discern the direction the process takes when the time scale is shorter and the region is smaller.

We can illustrate this "theory of mind–free" approach to history at a finer grain than Diamond's sweeping history. Consider the advent, persistence, and disappearance of foot binding in China. The practice sprang up a thousand years ago, persisted and spread throughout the most populous and in many ways the richest, most advanced civilization in the world, and then, after a millennium, unraveled in less than a decade, within the living memory of your great-grandparents. How did it happen? Explaining the emergence, persistence, and disappearance of this practice would seem to be a task for the historian and the cultural anthropologist. Working together, they might seek to identify the meaning of a practice for the Chinese family over forty or more generations, with its aesthetics and ethics articulated in the narratives of participants. But a far more compelling explanation emerges from the application of game theory, an explanation that identifies a mechanism of rapid emergence, long persistence, and sudden disappearance, all quite independent of the details of what any participant—father or daughter, suitor or in-law—wanted or desired or expected. It's not that participants in this practice didn't have beliefs and desires. Game theory, in particular evolutionary game theory, explains the millennium of Chinese foot binding without any need for a theory-of-mind narrative that makes use of people's actual motives, interests, beliefs, and expectations.

Scholars pretty much agree that foot binding first appeared in the imperial palace during the Sung dynasty roughly 1000 AD, spread from the court down through the class system and across the country, becoming more exaggerated over time, and affecting approximately two-thirds of women who were not required to work in agriculture, manufacturing, or the trades. Repeated imperial edicts forbidding the practice had no effect on its persistence, but then, in the first decade of the twentieth century, it suddenly ended.

Various cultural explanations for foot binding have been offered: it served as a racial or gender marker, enabling people to distinguish Chinese from non-Chinese women or women from men; it was believed to foster fertility and good health: it distinguished women who came from higher-status, wealthier families from women who did not; men believed foot-bound women made better wives, being both more modest in public and supposedly more wanton in private. Historians have found many records that the explicit purpose of foot binding was to confine women and enhance their morality and marital opportunities. Western theorists from Sigmund Freud to Thorstein Veblen to Marxian social scientists have offered a variety of explanations for foot binding. Unsurprisingly, each explanation appeals to the theorist's own distinctive theory, be it conspicuous consumption, castration anxiety, or patriarchal control of women's every movement.

Perhaps these explanations have some merit, but what they don't do is provide an explicit, unified account of why, once it emerged, foot binding (1) spread rapidly throughout China, (2) persisted for more than a thousand years despite its severe costs to many women and their families, and (3) suddenly unraveled completely.

As Gerald Mackie argues convincingly, evolutionary game theory explains all three of these developments in the history of foot binding in China (Mackie, 1996). They all arise through the operation of the same small set of factors governing the means-ends behavior of human beings over a period when the environmental conditions were changing. Mackie's key insight is that the environment that selected for adaptive behavior in individuals is shaped by the adaptive behavior of other individuals. In this respect, his explanation is like Diamond's, a Darwinian one, though it operates in a smaller region over a shorter length of time to account for a much more specific process.

Start with two general facts: (1) like other higher primates, people engage in the sort of means-ends behavior that makes mind reading reasonably effective in local environments; and (2) there is always strong selection—genetic or cultural—for locally successful child-rearing practices, and for practices that maximize the probability that the children the male is supporting are actually genetically his own. Natural selection has found many different ways among sexually reproducing animals of solving this "problem of the uncertainty of paternity," and cultural selection has found more for humans.

The cultural variant of foot binding was selected for in the polygamous Chinese court a thousand years ago because it helped solve the uncertainty problem for an emperor or other powerful males with many consorts who wished to be certain the children the consorts gave birth to were actually theirs. Foot binding almost certainly didn't emerge because anyone planned it with the explicit goal of reducing the uncertainty of paternity problem. But marking consorts in this way and controlling their movement had this effect, and so foot binding was selected for, from the time it first appeared.

The social environment created for others in the court by this practice among emperors and their immediate families selected for these lower-ranking officials to engage in the same practice since it would make the officials' daughters more likely to marry up into the higher ranks by reducing paternity uncertainty. The process rapidly and repeatedly reshaped the cultural environment: once introduced at the court, foot binding shaped behavior down the socioeconomic hierarchy, quickly spreading until it reached a level at which the immediate short-term costs of foot binding—forgone labor—were too heavy for the poorest families to bear. There the spread of foot binding came to an end.

It's easy to sketch the strategic interaction problem of lower-ranking males facing the choice of whether or not to foot bind their daughters once the highest-ranking, richest, most powerful males begin to foot bind their own daughters so as to make them more attractive to the emperor. A parent faced two alternatives: (1) diminish the daughter's prospects for marrying upward but spare her the extreme pain of foot binding, or (2) subject the daughter to that pain and make it very hard if not impossible for her to work but increase her prospects for upward marriage into a richer, more powerful family that would support larger numbers of her children. At the outset, when the practice was spreading down from the court to the lower classes, the payoff for the second alternative was clearly much higher. There was strong cultural selection for any family that could afford to adopt the practice.

It doesn't matter to this explanation whether parents were calculating the relative payoffs of these alternatives or responding to social pressure, imitating those with higher status, or adopting a practice they found aesthetically pleasing, morally required, or enjoined by tradition. So long as the practice enhanced the numbers of well-supported offspring (and the benefits for their families) for males who adopted it, it was going to be selected for, spread, and persist.

In fact, once initiated, the practice quickly spread throughout the country until it became so widespread that every family but the poorest was trapped in what game theorists call a "suboptimal local equilibrium." No family could cease the practice of foot binding without condemning its daughters to spinsterhood, but in the competition with every other family to "marry up" their daughters, no family was benefiting either, because every family was doing it. Getting off the merry-go-round was impossible. But every family was also worse off than it would have been if all families were to get off the merry-go-round together. Once the practice became pervasive, however, the high costs of foot binding were no longer outweighed by its competitive benefits. That's what made the local equilibrium *suboptimal*. There was another local equilibrium that would have been better for all families: no family would foot bind its daughters while all the families' daughters would still compete to marry up. The trouble was that no family could reach this superior equilibrium unless all the families moved to it at the same time. There could be no gradual process in which one after another family defected from the practice of foot binding. If foot binding was going to end, it would have to end suddenly, not gradually.

Something had to happen to destabilize the suboptimal local equilibrium at the beginning of the twentieth century, something that would move all families to a "tipping point" at which suddenly they could all coordinate moving away from foot binding *together*. Some historians and anthropologists suggested it was China's modernization, urbanization, and industrialization in the early twentieth century that unraveled foot binding as a practice. But,

in point of fact, the practice unraveled half a century before these three factors began to play a significant role in Chinese history. Rather, there had to
be some sudden environmental change for developments to reach a tipping
point. But what kind of sudden change? Mackie, the theorist who advanced
the game-theoretical explanation for the historical process, identifies a few
factors: the introduction of westernizing tastes, which disapproved of foot
binding, and especially the efforts of reforming Western missionaries, who
agitated against the practice in the countryside; and industrial changes that
excluded foot-bound women from income-producing work, an especially
important factor in unraveling the practice among families in the lowest
classes where its net benefits in marrying-up were smallest. Once a group
of families just large enough to escape the suboptimal local equilibrium did
so, the snowball effects were enough to rapidly put an end to the practice
across the whole country, something imperial edicts had been unable to do
for centuries.

Notice how this historical explanation works; it's rational choice operating without the need for actual rational choosers. As economists long ago
recognized, a market in which people compete just naturally selects for
winners—buyers and sellers who do better—or at least selects against losers
who do worse, whether or not any of the competitors know what they are
doing. Darwinian processes, whether biological or cultural, "seek" locally
optimal outcomes. In that sense, they are "rational." But neither biological
nor cultural selection ever finds globally optimal outcomes, even if there
are any to be found.

The success of a nongenetic Darwinian approach to human affairs, at the
broadest scale of Diamond's work or Mackie's narrower one, is not the only
reason we can be confident that natural selection is the real driver in world
history over millennia, continental history over centuries, national history
over decades, and local history over a year or so. In fact, much of the conventional narrative history, written by conventional narrative historians
since they started writing down stories 3,000 years ago, demonstrates that
the driving force of history is this Darwinian process operating through
human culture.

Think back to Henry Kissinger's story about the Congress of Vienna.
Talleyrand, Castlereagh, Metternich, and Tsar Alexander all had purposes
clearly in mind, ones that explained their actions, right? And so did Britain
and France and Prussia. Somehow, all the beliefs and desires of the individual

British, French, and Prussians who had a hand in the Vienna negotiations added up to the purposes of the countries they represented. And then there was the Congress itself, an event that lasted for months, and that itself had a purpose, a goal, an objective—the "Concert of Europe," as it came to be called afterward. That's human history as Kissinger in this case and narrative historians in general tell it: on every scale, there are purposes achieved or thwarted. Through all the revisionism, all the archival discoveries, all the reappraisals, and all the hindsight, the teleology is pervasive.

But the whole history of science from Newton onward tells us there is no purpose in nature, only the appearance of purpose, the overlay our minds spread across the domains of nature, whether physical, biological, or psychological. Neuroscience just completes the project of banishing purpose from nature, showing that its appearance is an illusion, although once a highly adaptive illusion and perhaps still indispensable in everyday life, but, for all that, not real, so not available to explain anything.

Recall that Darwin revealed the causal mechanism, the "machine behind the curtain," that produced in us the illusion of purpose in nature—blind variation and natural selection. Perhaps he should have called natural selection "environmental filtration" instead to emphasize the passivity of the process, its complete freedom from even the suggestion of purposeful "selection." But it's too late to change the name of the theory, even though calling it something like the "theory of environmental filtration" would have made Diamond's reliance on it to drive the process *Guns, Germs, and Steel* describes much clearer.

It needs to be reemphasized how science first banished purpose from the physical domain, and then revealed the detailed mechanisms that produced the illusion of purpose wherever that illusion appeared. That purpose has no explanatory role and indeed no place at all in the physical domain is abundantly clear from the most fundamental facts of physics. Until Darwin, however, there was no way to banish purpose from the biological domain as well. Toward the end of the eighteenth century, the German philosopher Immanuel Kant wrote, "There will never be a Newton for the blade of grass," meaning that purpose could never be surrendered in the life sciences since, without it, the evident means-ends economy of nature couldn't be explained. But twenty-five years later, in 1809, the "Newton of the blade of grass" was in fact born in Shropshire, England. Darwin would advance the Newtonian project, banishing purpose from the biological domain, while leaving the appearance of purpose, even as he explained it away. Contemporary molecular biology would fill in the details of Darwin's theory, showing how both the heredity and development of organic systems were simply the result of the interaction of macromolecules, and how these systems interacted with their environment through the same physical and chemical processes to produce adaptations that we still take for the fulfillment of purpose.

At first, it was expected that the program of scientific research that had banished purpose from the physical and biological domains would leave the theory of mind intact in the psychological domain. Here the theory told us that we had beliefs and desires and acted on them and that our pairings of beliefs and desires caused our actions. Purpose was thus expected to preserve its explanatory role when it came to human affairs because purposes were *inscribed* in both beliefs and desires. Our desires identified our purposes—our immediate and ultimate ends—and our beliefs identified the means by which we might fulfill those purposes.

Neuroscientists were supposed to show us how our pairings of beliefs and desires caused us to act: how the beliefs and desires in our minds were expressed with *content* in the neural circuits of our brains, and how the firing of these neural circuits brought about our movements and actions.

But when neuroscientists instead found that the neural circuits neither had content nor needed it to perform their functions, it became clear that there was no room either for beliefs and desires or for purpose in those circuits or in the brain itself, even when almost all of us persisted in attaching purpose to beliefs and desires, both ours and other people's. Most neuroscientists were not surprised by this outcome. The behaviorists had anticipated it two or three generations before them but just didn't have the tools that neuroscientists had to achieve it.

With that outcome firmly "in mind," let's go back to the Congress of Vienna. It had no purpose, and neither did the machinations of any of its participants. In fact, none of them—not Metternich, not Talleyrand, not Castlereagh, and not Tsar Alexander—came to the Congress with a purpose. There weren't and indeed aren't any purposes, any more than there were or are epicycles or phlogiston. There was and is the *appearance* of purpose, of course, all over the biological domain, to include the subdomain of human affairs. Not knowing this, Kissinger invoked "purposes" throughout his story: they were the ends that explained what happened in Vienna as well as the means the participants employed to attain those ends. His readers shared his ignorance about the absence of purpose from the participants' actions and shared the theory of mind that made purposes explanatory. So Kissinger and his readers, including Richard Nixon, thought they had some useful knowledge, something they could take away from his history of the Congress and apply to the Cold War between the United States and the Soviet Union.

What was it, if not purpose, that drove the behavior, actions, and events that led to the Concert of Europe, the subsequent century of peace (apart, of course, from the Crimean War, the Danish-German Seven Weeks' War, the Austro-Prussian War, the Franco-Prussian War, and a few others)? Well, within the brains of the principals, neural circuits were being wired together by the processes that Eric Kandel, John O'Keefe, and May-Britt and Edvard Moser first discovered (and that their successors will go on to describe in ever greater detail). These processes will almost certainly turn out to be cases of operant and Hebbian conditioning acting on neural circuits that were themselves built by the same processes that operated in the development, growth, and maturation of the brains of the principals at the Congress. These processes of blind variation and natural selection in the brain are the same processes that drive Jared Diamond's account of world history over the last 13,000 years. And when the top-down explanations Diamond and Mackie pursue join hands with the bottom-up explanations that neuroscientists continue to pursue, they're able to explain away the fallacies of narrative explanation.

But most of us have questions we want to ask. What really happened at the Congress of Vienna—and why? What were the consequences of the Congress for Europe as a whole? For particular countries? And for particular participants? The story Kissinger tells us might satisfy our curiosity, and it might lull us into an unwarranted feeling of confidence that we've at least some of the answers to our questions about the Congress. But if, as we now can see; a story simply can't answer the questions we want answered about the Congress, what can? Well, some of the questions have no answers and, more important, need none because they rest on mistaken presuppositions. A great deal of our historical curiosity is driven by *Homo sapiens'* long love affair with the theory of mind. Not only does it drive our curiosity; it also seduces us into thinking its answers to the questions it gets us to ask can help us cope with our future, or even with the future of the world itself.

Like the questions people asked for centuries on end in service of the dead-end Ptolemaic and phlogiston theories discussed in chapter 10, a lot of the questions historians ask about the Congress of Vienna are pointless, starting with the ones about what all the players wanted to get out of the Congress for themselves and for their countries, and what each of them believed the others wanted. And when historians ask what the meaning of the Congress of Vienna was for the nineteenth and twentieth centuries, they're asking how the Congress figured in the pairings of desires and beliefs in the minds of later statesmen, monarchs, and others. And these questions are also based on radically false, but hard-to-uproot presuppositions.

Histories can certainly provide correct, accurate chronologies—what happened when and where in human history. We can assume that the events in Kissinger's narrative of the Congress actually happened and in the order he presented. And we may agree that the Congress was followed by a century without at least a Europe-wide war. It's tempting to stitch the events of the Congress together into a teleology—a process aiming at attaining a goal, purpose, end—the Concert of Europe. It's even more tempting to stitch together the events in the lives of each individual statesman at the Congress into a process culminating in a succession of aimed-at outcomes such as Talleyrand's success at restoring defeated France to its former status as a great power. But if we know there are no purposes in natural history, including human history, we'll recognize that the best we can do by way of explaining what happened is to stitch the same events together into quite a different trajectory, a Darwinian one or perhaps one traced by following the neural firings in the brains of the ministers and plenipotentiaries.

If there was a Darwinian scenario at Vienna, it was invisible to participants and subsequent historians alike. That there was a process that ran through the brains of the participants is certain, but tracing it out won't answer the questions we want addressed. The right Darwinian account of the process at Vienna in the summer of 1815 almost certainly wouldn't stitch together the same events historian's chronologies identify as significant. And if a neuroscientist were to describe the events there, no historian would be able to recognize them.

But there is a way to approach questions of history we can actually answer and whose answers might actually help guide us in the future. Was it the Congress of Vienna that brought about the Concert of Europe and the long quasi-peace on the Continent that was only finally destroyed by the kaiser's "blank check" to Austria-Hungary in 1914? That's a question that might be answered, but not with stories about the actual thinking of Metternich, Castlereagh, Talleyrand, Tsar Alexander, and their successors. The best answer to the question of why the peace held and whether the Congress of Vienna had anything to do with why is probably to be found in "evolutionary game theory," which identifies the conditions under which stable equilibria emerge among strategies that interact with one another. Identifying the strategies in the policy of each of the European powers and showing that each produced the highest payoff in light of the strategies the other powers adopted—year by year—would explain the persistence of the equilibrium through the nineteenth century. It would explain the Concert of Europe. And it would do so without recourse to anyone's belief-desire pairings. All we need assume is that some process of natural selection is operating in human affairs that produces the appearance of purposes, ends, goals, aims, and, in this case, a peaceful stable equilibrium. Evolutionary game theory would also lead us to expect that, at some point, the equilibrium would break down. It would warn us that "arms races" are the principal causes of such breakdowns. In describing an arms race, evolutionary game theory posits the persistence of continual random mutation in behavior that eventually produces a novel strategy among one or more players in the equilibrium, a new strategy that, being more advantageous than any other strategy to that player, at least in the short run, is selected for. History, however, whether natural or human, is never driven by adaptive selection in the long run. The Darwinian process can "look ahead" no further than the next exogenous environmental change or novel variation in response to it. The novel strategy, in breaking out of the equilibrium, destabilizes and destroys it, and changes circumstances in ways that in turn select for further novel variant strategies by others to counter each novel strategy. Whence the label "arms race." But, in human history, the arms race is no mere metaphor.

Thus evolutionary game theory is one tool we can use to answer some of the questions narrative history poses. And though it won't tell us why the kaiser issued his "blank check," finally putting an end to the Concert of Europe, it will help us understand how matters stood in Europe in the decade before Wilhelm unraveled the hitherto optimal strategies that had together kept the peace for a century. Evolutionary game theory may even provide a basis on which to design institutions that preserve peace in the future by obstructing or limiting arms races—both metaphorical and literal.

If we understand why the nineteenth-century peace held in Europe, can we also credit the Congress of Vienna as the reason it did? Can we identify the Concert of Europe as the consequence of the Congress, as Kissinger does? Was he right for the wrong reasons?

Perhaps, if the only way the strategies that synergized to constitute the post-Congress equilibrium could have emerged was by the process that occurred over the months Congress met. If that was so, then we can credit the Congress with bringing about the peace. But it's crucial here to note several things about such a conclusion that an approach like Diamond's or Darwin's reveals.

First, evolutionary outcomes in nature are often "robust" and sometimes convergent. That is, Darwinian natural selection is a strong force and it operates on such a large amount of persistent mutation that it can shape the same adaptive outcome repeatedly and from widely different starting points. Living in the ocean constrains the mammals that came to live there to adopt many of the same solutions to "design problems" that the fish adopted before them. In human cultural evolution, the rate of mutation— the appearance of novel behaviors, institutions, technologies—is much more rapid. A Darwinian process starting from different points but driven by a serious environmental threat or opportunity will exploit the slightest variations to move in the direction of the same adaptive outcome. If the emergence of the concert of Europe in the nineteenth century was like this—an outcome fated by environmental constraints, then the Congress of Vienna was by no means indispensable. If it hadn't happened, some other set of events would have driven Europe to the same outcome. Sometimes, there really is such a thing as "historical inevitability."

Second, there are many processes of natural selection going on in nature all the time, at different levels of organization. Darwinian processes at the level of genes, cells, organisms, kin groups or whole populations are all going on simultaneously, producing outcomes that aggregate the results of all these processes. The same is happening in human affairs: with the right data, we can trace out the selective environment in Vienna that resulted even in the social outcomes—who danced and slept with whom—as well as in the political outcomes that might have aggregated the effect of these social processes together with economic and military ones. Even if the Congress of Vienna wasn't a necessary condition for the Concert of Europe, it was itself still part of a Darwinian process. We know this because so many of its features show the appearance of purpose, and there is only one way the appearance of purpose can arise—the Darwinian way.

Third, it's unlikely but possible that the Congress of Vienna may have been more like the comet that crashed into Earth 65 million years ago and suddenly changed everything, killing off the dinosaurs and ushering in the age of mammals. It's certain that some historical events play this role, especially unpredictable technological or economic events: the invention of the stirrup, the emergence of nuclear weapons, and the stock market crash of 1929 may figure in evolutionary processes in this way. Think of the impact of the crash and subsequent economic depression on political outcomes in the United States, Great Britain, and Germany in the decade that followed. Then there are surprise attacks, political assassinations, unique, unpredicted, unpredictable events that disrupt equilibria or accelerate and shift evolutionary change. Kaiser Wilhelm sent Lenin through Germany and Sweden to St. Petersburg in the spring of 1917, just before the U.S. entry into World War One. The effects of this single act by the kaiser on the rest of the twentieth century were considerable. There was strong selection for such an act. The Germans were casting about for stratagems that would force the new Russian republic out of the war. But the evolutionary effect of Lenin's return to Russia on the rest of the twentieth century was rather more like the cometary impact on the Earth 65 million years previous.

It's worth pausing to note how different real historical explanations will be from the narrative ones driven by the theory of mind. Consider the twelve-person jury system at the heart of courtroom trials in the countries that observe English common law. Why does it still exist? Why twelve jurors and not eleven or thirteen? Why the particular responsibility of a lay jury to make findings of facts, but not of law. The conventional history tells a story, which starts sometimes with the Viking invasion of Britain and sometimes with King Henry II's decision to establish a system investing twelve men of a jurisdiction to adjudicate land disputes. The story goes on to invoke trial by ordeal for those whom such a jury reliably reported to have committed a crime. Then, in 1215, the ecclesiastical authorities decided to order clergy to withdraw from trials by ordeal, and jury members began to render verdicts instead of reporting suspicions. In the same year King John believed that to survive he had to grant the barons Magna Carta, which established the right to jury trials by peers of the defendant. Narrative history continues to trace the process whereby the jury system and the number of jurors persisted through centuries of practice, citing statute law, royal decree, and conflicts over the king's unchecked powers, especially the "star chamber," to the point at which the practice became so well entrenched that no readers interested in the matter need their curiosity further satisfied. Much of the history is the citation of texts—Acts of Parliament, legal writing, judicial opinions, and, of course, stories of jury findings overthrown, disregarded, reinstated, and treated as precedents. Scrupulous historians of the law have long been at pains to establish the facts, the observable events in this history, but they have never questioned how the theory of mind holds the links in the chain of their history together.

The approach that will actually explain, instead of simply satisfying feelings of curiosity and scratching our conspiratorial itch, begins with many of the same "data points" that narrative history uses, all the way back to King Aethelred's Wantage laws in the tenth century. But then it searches for circumstances in the historical environment of the juries that selected for the persistence of the institution. In some cases, it will find them—the growth of baronial power and the weakness of King John at the time of the signing of the Magna Carta in 1215, for example. It may have to treat as a coincidence the fact that the ecclesiastical authorities withdrew the clergy from trial by ordeal in the same year that the Magna Carta was forced upon King John. The search for the real forces that shaped English common law and established the jury system will help itself to factors and forces quite beyond the explanatory resources of the theory of mind. For example, the explanation of why the jury system emerged, persisted, and strengthened will cite the need of any society to find ways to peacefully adjudicate conflicts and will be able to employ a comparative method—examining the impact of different practices in the same or similar societies—to decide which practice is most adaptive, meeting this need at minimum cost. It may treat features of the jury system as "frozen accidents," which have no interesting explanation at all. For example, why twelve jurors and not six or fifteen, eleven or thirteen? The reason may simply be that Henry II chose an arbitrary number. Or it may be that there was a natural experiment in which the number of jurors was varied over time with outcomes that were better or worse at meeting social needs and conferring benefits or at advantaging those in a position to enforce their authority. Starting from the same data, the real historical explanation, however it proceeds, will be completely different from the one underwritten by the theory of mind.

Why can't the explanations of both the Darwinian and the theory-of-mind approach be true? Why can't they be combined to produce an even more powerful explanation? Simple. The theory of mind–driven explanation doesn't add anything to the real (Darwinian) explanation because not a single one of the links it cites was actually forged by the historical process. Of course, underlying the real process of behavioral variation and cultural natural selection, there were incalculable numbers of neural events going on in the brains of the vast number of people who brought about and then ensured the persistence and spread of the jury system. But these brain events weren't anything like the belief-desire pairings the theory of mind–driven narrative explanation needs.

Imagine a phlogiston chemist insisting that Antoine Lavoisier's explanation of combustion would be improved if Lavoisier could only add to oxygen's role a role for phlogiston, too. Silly, right? (Although probably not to those who still believed in phlogiston after Lavoisier's discoveries.) The difference between adding phlogiston and adding narrative history to the mix is that we're wedded to the theory of mind that drives narrative history in a way we no longer are to the phlogiston theory. We just feel that narrative history's explanation makes a real contribution to our understanding —and that, in any event, it can't hurt; it can only help.

There's a better comparison that will help us see the mistake of trying to combine narrative histories with the real historical process. People have repeatedly tried to insert God into Darwinian processes of blind variation and natural selection ever since 1859. Theists, in particular, have sought to reconcile Darwin's processes with God's by attributing to God the power both to create the variations that met the needs of, benefited, or conferred advantages on species and to select among them accordingly. Combining Darwinian with divine processes would have comforted believers generally and would have avoided conflicts for biologists, many of whom are also believers. And it might have made Darwin's theory of blind variation and natural selection easier to accept for everyone. Notice that, by invoking God's benevolence, omniscience, and omnipotence, it would even have harnessed together Darwin's theory and the theory of mind. The trouble with such an effort is obvious. It's not just that adding God would add no predictive power to Darwin's theory. We know, and Darwin knew, that all variations are chancy and most variations (mutations) are harmful, and that many advantageous mutations are eliminated by random forces before they can make an evolutionary contribution. And, of course, subsequent developments in genetics explained both of these phenomena and provided a tested and confirmed scientific explanation of how variations actually arise. No role for God's beneficence, omniscience, and omnipotence here. In fact, invoking God's powers and attributes actually distracts us from real understanding in the biological domain.

Exactly the same goes for the real understanding of human history. Not only does the theory of mind add nothing by way of predictive power to a Darwinian approach to human history. Developments in science since Darwin, especially in neuroscience, have shown us why it doesn't. And accepting the theory of mind on its own terms throws up serious obstacles to further scientific developments, obstacles at least some philosophers are quick to point out (Churchland, 1986; Stich, 1983). Nevertheless, since the theory is bred in the bone, we have even more trouble shaking it than we do belief in God.

Even Jared Diamond is not immune to the theory's charms, leaving at least a little room for narrative explanations in his *Guns, Germs, and Steel*. Most people, including Diamond, will explain the occasional historical contingencies that his theory requires by invoking the theory of mind. Why did the eunuchs at the Chinese court lose to the Mandarins? Why did Ferdinand and Isabella agree to fund Columbus's expedition even when four other sovereigns had refused to? The answer must be something about what the principals believed and wanted. How do we know? Here's the last paragraph of Diamond's book:

> Successful methodologies for analyzing historical problems have been worked out in several fields. As a result, the histories of dinosaurs, nebulas, and glaciers are generally acknowledged to belong to fields of science rather than the humanities. *But introspection gives us far more insight into the ways of other humans than into those of the dinosaurs*. I am optimistic that historical studies of human societies can be pursued as scientifically as studies of dinosaurs—and with profit to our own society today, by teaching us what shaped the modern world, and what might shape its future. (p. 425: emphasis added)

What is Diamond telling us here? That once successful methodologies for studying some domain are worked out, the domain is taken away from the humanities and becomes the preserve of science? It's clear from the immediately preceding paragraphs that Diamond's "successful methodologies" are those which employ the experimental method either in the lab or by identifying "natural experiments"—"comparing systems differing in the presence or absence (or in the strong or weak effect) of some putative causal factor" (p. 424). You'd think Diamond was going to go on in this last paragraph to tell us that, once we discover the right methodology, we'll be able to take the history of humans away from the humanities and make it the subject of explanations tested and confirmed by empirical science. But no. Suddenly, in a sentence that doesn't fit in anywhere in the 425 pages that come before it, he switches directions completely and gives pride of place to "introspection," of all things, as a source of knowledge about humans.

It calls to mind the change Darwin made to his sixth and final edition of *On the Origin of Species* in an attempt to avoid further controversy. In the five previous editions, after some 500 pages of perhaps the most powerful argument ever crafted for the effective absence of God from the workings of evolution, Darwin ended with a beautiful peroration on natural selection:

> There is grandeur in this view of life, with its several powers, having been originally breathed into a few forms or into one; and that, whilst this planet has gone cycling on according to the fixed law of gravity, from so simple a beginning endless forms most beautiful and most wonderful have been, and are being evolved.

Then, in 1872, twenty-three years after the first edition, Darwin inserted the words "by the Creator" after "breathed" in "having been originally breathed into a few forms or into one."

Diamond's about-face amounts to a cop-out of the same magnitude as Darwin's. He knows full well that introspection can never operate or be tested by controlled or natural experiments. He knows that in the millennia before *Guns, Germs, and Steel*, historians had been offering answers to his very questions by using their powers of introspection, with spectacularly unreliable results for our future guidance.

As he writes in the very last sentence of his book—just after invoking "introspection"—Diamond is optimistic about the scientific study of human history and its potential to guide and ameliorate the human condition. But such study will require him and the rest of us to turn our backs on introspection, give up the theory of mind, and consign narrative explanation to the creative arts.

13 *The Gulag Archipelago* and the Uses of History

In 1973, Alexandr Solzhenitsyn published the first of his three-volume narrative history of the Soviet Gulag *cum* autobiography of his life as a "zek" (inmate) in the system of forced labor prison camps that spread like a chain of islands across the eleven time zones of the Soviet Union. The word "Gulag," originally an acronym for the authority that ran these prison camps, has entered most languages as a proper noun meaning any political prison extracting forced labor from its inmates. Solzhenitsyn's aim in *The Gulag Archipelago* was not only to indict Joseph Stalin, his henchmen, and, before them, Vladimir Lenin for the crime of creating and administering a system that killed one and three-quarter million people (Solzhenitsyn, 1973–1978). His aim was also to show that the Gulag was an inevitable outcome of the mind-set their October Revolution gave rise to. The Gulag was the unavoidable means to their ends; both men were inseparably harnessed to it by the Bolsheviks' project.

Naturally, Solzhenitsyn was unable to publish the book in the Soviet Union. In fact, he had to keep the entire project secret for years. One of those who copied the manuscript hanged herself after revealing its location to the KGB. Once published in the West, however, *The Gulag Archipelago* was translated into more than thirty languages and became an international best seller. Three years before, in 1970, Solzhenitsyn had won the Nobel Prize in Literature for other works, which had also been suppressed in the Soviet Union. The worldwide acclaim he received meant he couldn't be jailed again without serious repercussions; instead, in 1974, Solzhenitsyn was expelled to the West, where he would live for the next twenty years.

That *The Gulag Archipelago* had a major effect on the subsequent life span of the Soviet Union is hard to deny. It was evident that the KGB, the

Politburo, and Communist Party hierarchy in Moscow thought so, evident, that is, if you can believe the theory of mind. They did everything they could to suppress the circulation of smuggled-in and "samizdat" (self-published) copies of the banned work and to counter its effects, even at great and evident cost to themselves and their interests.

Certainly, Solzhenitsyn's narrative history had a greater impact than his Nobel Prize–winning fiction, his magnificent novels of Soviet life, *The Cancer Ward*, and *The First Circle*, both published in the West in 1968.

How much was done to unravel the Soviet Union by Solzhenitsyn's three-volume narrative history of the Soviet Gulag *cum* autobiography of his life as a *zek* between its publication and the end of Communism in the Soviet Union? Some might cite the election of the Polish Pope John Paul II in 1978 as making a greater difference. Others might point to U.S. President Ronald Reagan's "Star Wars" or Strategic Defense Initiative or to Soviet President Mikhail Gorbachev's policies of glasnost and perestroika,—or perhaps to Gorbachev's wisdom, or his heroism, or rather his fecklessness as the decisive factor.

The trouble with narrative history is that it simply can't resolve these questions one way or the other. As we saw in chapter 2, there are too many forces operating on the trajectory of human affairs even to be enumerated and that weighting them in any one individual case, let alone in more than one over time, is a fool's errand. Certainly, there's nothing in narrative history that would enable us to accurately weight the causal factors that led to the end of the Soviet Union in 1991. But what we can't deny is that *The Gulag Archipelago* had a profound effect on people everywhere and on late twentieth-century events—not only in the Soviet Union and its successor state, the Russian Federation, but elsewhere as well.

The reason is obvious: it moved people; it had an enormous impact on people's emotions and motivations. It provoked feelings—of anger and hostility, sadness and despair, outrage and revenge. It also provoked feelings of admiration for Solzhenitsyn and for those whose stories he told, even exhilaration at the courage and resilience of the human spirit in the face of adversity. As the three volumes appeared from 1973 to 1978, they provoked people to action, to buying and reading them, and to talking, arguing, and writing about them. But they also provoked political activity and significant change in the values, especially among people of the Left, who found themselves surrendering illusions and forsaking commitments that had been among their most cherished.

Great works of history, especially of narrative history, have this emotional impact on us. They make a difference in our lives that is often much greater than the difference the greatest works of historical fiction make. *The Cancer Ward* and *The First Circle* did not seem to contribute much to the unraveling of the Soviet state, whereas *The Gulag Archipelago* clearly did.

Cognitive social psychology has a pretty good handle on how and why narratives move people to action by their impact on emotions. Perhaps the best understood emotions are fear and anger.

As noted briefly in chapter 4, some emotional responses are close to hardwired, especially, for example, fear of spiders and snakes, in whose presence it takes most infants almost no experience at all to feel the emotion of fear, an evolutionary adaptation selected for over millions of years of hominin experience in sub-Saharan Africa, where dangerous spiders and poisonous snakes were endemic (Öhman and Minetka, 2001). As the emotion is aversive and unpleasant, fear brings about a highly adaptive flight response. Other emotional responses are shaped by the environment during early development. "Anger" labels the emotional response whose display is generally reinforced in people from childhood onward when they are subject to treatment they and others deem as failures to reciprocate in games or small acts of disloyalty. In most normal adults, this response has been shaped into one felt when they experience injustice or unfairness or observe it inflicted on others (Henrich et al. 2004; Fehr et al., 2012). It's easy to see how the very same facts shaping cooperation and coordination on the African savanna would exploit the availability of an emotional response to failures of reciprocation to strengthen cooperation and coordination in the face of the challenges hominins faced. In fact, exploiting the emotion of anger at unfairness might have been indispensable to the enforcement of cooperation. Strategic interactions between subjects in controlled experiments show that opportunities to impose punishment on free riders substantially strengthens cooperative behavior. The same experiments show that in the absence of punishment cooperative behavior breaks down even when its continuation is to the benefit of all participants, who invariably identify feelings of anger at free riders as the motivation for acts of punishment (Fehr et al., 2012). Other studies show that people generally express more satisfaction watching or hearing stories in which offenders are punished rather than forgiven (Haidt and Sabatini, 2000).

Our conscious awareness of feelings of anger is followed by movement, behavior, action. We no more understand how this happens than we understand why consciously willing an arm to rise is followed the arm's doing so. Our ignorance of how this happens is actually disguised from us by the sensory data we mistakenly treat as giving the content of "willing an arm to go up." The same goes for anger. But fMRI data on emotions help us understand how an emotion such as anger leads to certain behavior. Even when no action ensues from the conscious feeling of anger, neural circuits that fire on such an occasion include portions of the motor cortex. Anger is a brain state constituted not just by the felt emotion but also by the beginnings of a process of bodily movement (Hortensius et al., 2016).

Cooperation-enforcing anger is an emotional response that relies on and requires the normal working of our mind-reading ability (Frank, 1988). That much is obvious. In the Pleistocene, when our ancestors would observe their fellow hunters or gatherers failing to discharge their roles, it would have been important to respond appropriately to their failure. Would anger have been an appropriate response? Anticipating a failure and preventing it would have been equally important; feeling and expressing anger to do this may have been highly adaptive. Once mind reading and language together produced the theory of mind, very specific linguistic content would have been attached to the emotion of anger, something like "I was just furious that he wouldn't carry his fair share of the load."

Narrative history's stories thus have an important impact on us, owing to their role in provoking our emotional responses, especially the responses that lead us to action. Although, in this respect, narrative history is no different from other cultural artifacts that move us—art, music, drama, narrative fiction—it's often much more effective than they are in moving us to action. *The Gulag Archipelago* shaped the history of the last twenty-five years of the twentieth century owing to its impact on people's emotions, especially their anger at its portrayal of injustice on a vast scale.

Alas, the same thing can be said of Adolf Hitler's *Mein Kampf*. It, too, mobilized millions of people to engage in atrocities unrivaled by anything except those ordered by Stalin. Unlike Solzhenitsyn's *The Gulag Archipelago*, Hitler's *Mein Kampf* is tough sledding. I suspect hardly anyone, not even the most ardent Nazi, was able to read it all. The book is a farrago of lies about

Hitler's own life combined with hundreds of pages that seem to scream with Hitler's hard-to-forget voice (as preserved in newsreels, documentaries, and propaganda films) about history, some of it verifiable, but most of it just made-up narrative explanations, easy enough to understand (although just as easy to disbelieve since the "evidence" Hitler offered was laughable).

There are vast differences between these two. One is the work of a great writer reflecting decades of careful research, a factual chronicle of profound moral force. The other is the illogical, disconnected scribbling of a madman, a disorganized screed of hateful fiction masquerading as history and implicated in the death of millions. But both proceed by employing the same explanatory theory: their stories are driven by the theory of mind. And both changed the world in profound ways because their readers held both to be true. It's tempting to suggest that if they'd been read as simple works of fiction, they would surely not have had the impact they did.

But this confidence in the greater power of recognized or believed truth to move us in ways falsehood and fiction cannot would probably not be justified. Take, for example, two famous works of fiction that were both highly popular and extremely influential in their day—Harriet Beecher Stowe's 1852 anti-slavery novel *Uncle Tom's Cabin* and D. W. Griffith's 1915 pro–Ku Klux Klan film *The Birth of a Nation*.

It's commonly reported that when Stowe was presented to Abraham Lincoln, the president said, "So this is the little lady who started this great war." *Uncle Tom's Cabin* sold more copies in the nineteenth century than any other book but the Bible, and the anger that it provoked toward slavery was evidently every bit as fervent as the anger that *The Gulag Archipelago* provoked toward Soviet Communism. For its part, Griffith's film *The Birth of a Nation* is widely recognized to have greatly helped fuel the revival of the Ku Klux Klan in the 1920s.

What both these works of fiction share with narrative history is, of course, their narrative structure, the stories they tell, and the theory of mind they rely on in doing this. But they are fiction, and they were intended as such, received as such, treated as such. What this reflects is obvious: narrative alone, all by itself, can move us to action.

The emotive and affective power of narrative history and fiction masquerading or mistaken for narrative history is hard to overstate. Especially since the rise of romantic nationalism in the nineteenth century, ethnic, linguistic, and religious groups have been mobilized by their proprietary narrative histories of mistreatment, discrimination, even genocidal suppression into the mistreatment, discrimination, and genocidal suppression of other ethnic, linguistic, and religious groups. Politicians are far from alone in employing narrative history to appeal to their followers' sense of grievance or destiny. Many great creative artists and, of course, not a few sincere and scrupulous historians have done so as well. Two hundred years of nationalism that made the preservation of "historical memory" into a moral imperative are at the source of most of the violent episodes in history since the Congress of Vienna.

It was only in 2016 that someone finally noticed that this use of narrative history might not be such a good thing. *In Praise of Forgetting* is David Rieff's reflection on the contemporary "moral authority" of narrative history:

> [M]ost arguments in support of collective memory as a moral and social imperative ... seem to take as their point of departure George Santayana's far too celebrated false injunction, "Those who cannot remember the past are condemned to repeat it." This is the view that has become the conventional wisdom today, and the conviction that memory is a species of morality now stands as one of the more unassailable pieties of our age.... To remember is to be responsible—to truth, to history, to one's country, to the traditions of one's people or gender or sexuality (in this last instance what is usually meant is a group' suffering, the history of its oppression). Anything less is an act of irresponsibility that threatens to undermine both one's community and, in our therapeutic age, oneself as well. And even to question this consensus is to disturb what Tony Judt, writing in praise of Hannah Arendt, described as "the easy peace of received opinion." (Rieff, 2016, pp. 58–59)

Rieff's book identifies some of the motivation of the search for histories that have been made invisible by the hegemony of the West (or men or white people or Christians or straights). Some of the marginalized histories are laudable enough, well intentioned, indeed admirable. Some seem indispensable to prevent repeated genocides. Others aim at reconciliation: truth and justice. But too many of these marginalized histories have inflamed the irredentism and revanchism of the twentieth century and unabated continue to do so into the twenty-first. How much ethnic cleansing—of the Greeks and Armenians by the Turks, of the Hutus by the Tutsis (and vice versa), of the Croatian Muslims and the Kosovars by the Serbs, of the Arabs by the Israelis, was justified by such histories? "The crucial point is this," writes Rieff, we do not have to deny the value of memory to insist that the historical record (the verifiable one, not the mythopoeic one) does not justify the moral free pass that remembrance is usually accorded today. Collective historical memory and the forms of remembrance that are its most common expression are neither factual, nor proportional, nor stable. To be sure, were the political implications of this largely positive, or failing that, at least largely neutral, then arguing for a more skeptical view of remembrance would be both disrespectful to all those people to whom it provides strength and solace, and unnecessary. But is this the case? (p. 36)

The situation is actually much worse than Rieff's rhetorical question suggests. It's not just that "collective historical memory" is "neither factual, nor proportional, nor stable." The same problems arise in the best, most disinterested, archivally scrupulous, primary source–driven historical scholarship. If the historical record is anything more than a chronology, it's not verifiable. It's wrong. And wrong in the most dangerous way, the way pretty much guaranteed to ensure that the mayhem of the last 5,000 years of recorded history will continue into the future.

Narrative history is not verifiable because it attributes causal responsibility for the historical record to factors inaccessible to the historian. And they're inaccessible because they don't exist. The causal factors narrative history invokes—the contentful beliefs and desires that are supposed to drive human actions—have all the reality of phlogiston or epicycles. So narrative history, even at its best, is just wrong about almost everything besides the chronologies it reports.

But, as Rieff realizes, it's also dangerous. What makes "the historical record" presented by narrative histories dangerous is obvious, at least from the perspective of the cognitive sciences: reliance on the theory of mind is what makes narrative histories breed the emotions that have wreaked the havoc of recorded history: anger, shame, jealousy, retribution, vengeance.

By now, it's obvious why narrative history is a poor guide to the future, why its track record for providing usable knowledge that actually enables people to cope with their futures is abysmal. Matters are far worse, however. Not only has historical storytelling led us astray in our expectations about the future. It has more often than not led those who believe it into moral catastrophes. No one can seriously suggest that, on balance, narrative history has been a force for good since it began to be written down some 5,000 years ago. There are, of course, exceptions to this woeful track record, ones we honor even as we try to overlook the atrocities sincerely perpetrated in the name of history.

Here again it's worth distinguishing between the theory of mind and mind reading as an ability we share with other primates and, indeed, most other mammals. The ability to mind read is a skill hominins deployed face-to-face long before *Homo sapiens* emerged. It was responsible first for their and then for our survival, if only by making possible the domestic division of labor, mutual protection, and the teamwork of hunting and gathering. Mind reading has certainly been harnessed by the prosocial emotions of love, sympathy, and empathy right down to the present day. Indeed, in the history of our species, the positive uses of mind reading may, for all we know, far outnumber the harmful uses to which the theory of mind has been put. It's impossible, of course, to tot up the uses of each and decide.

But mind reading becomes malevolent when the theory of mind enables it to unleash our hostile emotions against people we don't even know, people who may even be long dead or far away.

Teleological "thinking" about—divining purposes in—the past was narrative history's bane for as long as it was its raison d'être. Darwin banished purpose from biology just as rigorously for humans as he did for other animals. But no one noticed. Academic historians sought with great success to drive teleology out of their discipline, though not for the right reasons. They rejected the notion that history was going anywhere because they rejected the Christian, Muslim, Marxian, capitalist, racial, patriarchal, and nationalist eschatologies that identified history's end, goal, or purpose.

Most people, especially those who drag narrative history into politics, didn't get the message. National narratives especially move people by giving meaning to the chronicle of their history. People mistake the emotions such narratives foster for understanding. When the broad sweep of a narrative history comes packaged in a story—Manifest Destiny, The White Man's Burden, "the civilizing mission [*la mission civilatrice*]," "blood and soil [*Blud und Boden*]"—it's hard for anyone to shake it just because of the way it's packaged.

The search for meaning in particular episodes, eras, or epochs in national narratives is driven by this teleological mistake. It takes its inspiration from the meaning people find in their own actions and in the stories they tell about why they acted. National narratives imbue groups, classes, sects, races, peoples, cultures, and movements with the same kind of motives and values, the same kind of recognition that ends require the means that the theory of mind accords to individuals. These national narratives are as hard to shake as a good biography or a powerful work of literature is. This book has sought to show exactly why such stories are addictive.

One tipoff that these master narratives are driven by the same machinery, the same theory of mind we employ in everyday life can be found in the morals so often drawn from them. Almost every national narrative underwrites a claim of rights or an accusation of responsibility, and these are collective, group rights and collective, group wrongs. But there's only one way such judgments make sense, and it's also the only way history in the form of a national narrative can justify them. Only agents acting on motives—good or evil—and beliefs—right or wrong—can be praised or blamed for outcomes. So either it's the nation as a whole acting in ways the theory of mind informs us about, or it's individual people who are, thereby defending their nation's rights, committing their nation's wrongs on others—or both.

In the nineteenth century, the first of these two alternatives was seductive. Under the influence of the German philosopher Georg Wilhelm Hegel, nationalist idealists asserted that a "world mind" (from the German *Weltgeist*) invisibly controlled the path of history with the emergence of the nation-states (of Europe especially). In asserting this, they were only articulating an idea that comes naturally to hyperactive agency detectors like us.

But absolute idealism went out of fashion in philosophy, and twentieth-century international jurisprudence rejected "collective responsibility" just in time to try individual Nazis at Nuremberg. Still, the personification of the nation, the people, the race, the culture has persisted, thanks largely to demagogues.

If there is meaning in history, if national narratives have significance and a moral for the future, it can only be through the meaning and significance of the acts of the individual people that drive narrative history. But, as neuroscience reveals, there's nothing that can do the work required to give people's actions the meaning or purpose that narrative history—whether national or personal—requires.

What should we rely on to cope with the future if not narrative history? We need only look back to chapter 12 to find the resources we require. They're the same resources we employ to cope with the biological, climatological, ecological, agricultural, demographic, and medical future. We need only figure out how to apply the tools that, with ever increasing success, have enabled us to cope with nature to our psychological, social, economic and political futures.

Stories are for children and for the child in us all. Nothing will ever stop us from loving them, at least not until natural selection radically changes our neurology. Narrative historians, like other storytellers, will never want for an audience. But we will all benefit by recognizing what narrative history at its best and most harmless actually gives us—not knowledge or wisdom, but entertainment, escape, abiding pleasure. That should be enough.

The Back(non)story

Nowadays, it's startlingly easy for readers with laptops (or desktops, for that matter) to check up on the claims of an author, provided they get to the right place online. By and large, Wikipedia is highly reliable. In fact, measured in a variety of ways, it's more reliable and certainly more up to date than standard reference works of previous generations, like the venerable *Encyclopædia Britannica*. If you don't believe me, check out the article "Reliability of Wikipedia" on Wikipedia, https://en.wikipedia.org/wiki/Reliability_of_Wikipedia. There are also highly specialized Internet sites that have become the standard reference sources in many disciplines. In philosophy, for instance, the *Stanford Encyclopedia*, https://plato.stanford.edu/, is beyond reproach in the distinction of its authors, the accuracy, completeness, currency, and scholarly references of its articles.

Sometimes a lot of footnotes, endnotes, references at the back of a book can instill a false sense of confidence in its readers, who will naturally suppose that the author's guidance is the best authority that can be consulted. So the following brief guide to the literature behind the claims of this book will, I hope, encourage readers to browse, search, re-search, update, and, most of all, move deeper into the matters introduced in the preceding pages. All you'll need is a knack for expressing the right question and an Internet connection.

Chapter 1: Besotted by Stories

Little need be said about the attractions of stories told in the histories and biographies that are published weekly in every language. The ones mentioned in this chapter appeared in the weeks before the chapter was first drafted. Had I waited a few more weeks, I would have offered different examples. A continually updated sample of the sort of nonfiction writing that seeks to convey understanding by telling nonfiction stories can be found on the History Book Club website, https://historybookclub.com.

Chapter 1 also identifies some much more enduring best sellers that seek to convey knowledge of science and technology by stories—histories and biographies, Bill Bryson's *A Short History of Nearly Everything* (Bryson, 2008) and Stephen Hawking's *A Brief History of Time* (Hawking, 1988). It will be easy for readers to identify many more such books that tell stories to convey or to substitute for the theories, models, laws, and mathematical formulas in which scientists express their achievements to one another. Just consult the Library of Science Book Club website, https://www.libraryofscience.net, which keeps book club members abreast of the latest in popular science storytelling.

What's hard to find are those rare and wonderful works of science writing that don't substitute easy-to-remember plotted narratives for the story-free results that scientists themselves seek. Brian Greene's *The Elegant Universe* (Greene, 1999) and *The Fabric of the Cosmos* (Greene, 2004) and Sean Carroll's, *From Eternity to Here* (Carroll, 2010) stand out as examples of such works in physics, as do several of Richard Dawkins's works in biology, his *The Selfish Gene* (Dawkins, 1976) among many others, and Steven Pinker's *How the Mind Works* (Pinker, 1997) and *The Stuff of Thought* (Pinker, 2007) in psychology.

There are, of course, important, indeed, even briefly popular books about history, historical trends, and even historical events that convey real understanding. They do so largely without appealing to the sort of narratives, plots, stories, lives—great and small—that hold general readers' attention. A good example is Thomas Piketty's *Capital in the Twenty-First Century* (Piketty, 2014), which explains the trajectory of economic inequality over a several-hundred-year history. It does so not with stories, but largely by appeal to a disarmingly simple inequality, $r > g$, which states the long-term rate of return on capital exceeds (is greater than) the long-term growth rate of the economy. Many people have bought this book, although few have read it, claiming that the narrative "drags."

Chapter 2: How Many Times Can the German Army Play the Same Trick?

If you're interested in learning more about America's participation in the First World War, the U.S. Army Center for Military History website, provides online access to the vast histories of it, https://history.army.mil/html

/bookshelves/resmat/WWI.html. The United Kingdom National Archives does the same for Britain's participation in the war, http://www.national archives.gov.uk/help-with-your-research/research-guides/british-army -operations-first-world-war/.

It's easy to find Barbara Tuchman's *The Guns of August* (Tuchman, 1962), John Eisenhower's *To Lose a Battle* (Eisenhower, 1969), Alastair Horne's *The Bitter Wood* (Horne, 1969), and all their successors online at Amazon.com. Milestones, especially centenaries, trigger multiple publications. In 2014, there were at least half a dozen new books seeking to explain the origins and the trigger for the First World War.

James McPherson reflects on the meaning of the American Civil War for generations of Americans in his *The Battle Cry of Freedom* (McPherson, 1988) and *This Mighty Scourge* (McPherson, 2009).

Chapter 3: Why Ever Did Hitler Declare War on the United States? That's Easy, Too Easy

Philosophers of psychology have pretty much agreed on the broad outlines of the diagram for the theory of mind in figure 3.2. The particular version illustrated there comes from the work of Shawn Nichols and colleagues, "Varieties of Off-Line Simulation" (Nichols et al., 1996), available on the web at http://cogprints.org/376/1/sim3.html.

The theory of mind as we employ it has been given other labels by philosophers for everyone's implicit use of it: "folk psychology," "common-sense psychology," "the theory theory, "the intentional stance," among others. Important early and influential contributions to work on the theory include Daniel Dennett's *The Intentional Stance* (Dennett, 1987), Jerry Fodor's *The Language of Thought* (Fodor, 1975), and *The Modularity of Mind: An Essay on Faculty Psychology* (Fodor, 1983). Alvin Goldman's *Simulating Minds: The Philosophy, Psychology, and Neuroscience of Mindreading* (Goldman, 2006); Kim Sterelny's *Thought in a Hostile World: The Evolution of Cognition* (Sterelny, 2003), and Steven Stich's *From Folk Psychology to Cognitive Science* (Stich, 1983).

A great place to start exploring this literature is Ian Ravenscroft's article "Folk Psychology as a Theory" in the *Stanford Encyclopedia of Philosophy* (Ravenscroft, 1997; rev. 2016).

Chapter 4: Is the Theory of Mind Wired In?

The connection between autism and the theory of mind was famously first advanced by Simon Baron-Cohen in his *Mindblindness: An Essay on Autism and the Theory of Mind* (Baron-Cohen, 1995). Baron-Cohen has also argued for the innateness of theory of mind. Important techniques that revealed the onset of theory of mind–driven behavior in infancy and early childhood were developed by Andrew Meltzoff, who later collaborated with Alison Gopnik, another significant figure in this research, and Patricia Kuhl to write *The Scientist in the Crib: What Early Learning Tells Us about the Mind* (Meltzoff, Gopnik, and Kuhl, 1999).

Thirty years of twin studies have convinced most researchers of autism's genetic basis. But as with other such traits, a large number of candidate genes—twenty-five or more—have been identified as linked (correlated) with the trait. A short and accessible introduction can be found at the Tech Museum of Innovation (Stanford University) website, http://genetics .thetech.org/original_news/news49/.

A widely viewed introduction to the fMRI techniques that localized the theory of mind to specific brain regions is a TED talk by one of its most prominent developers, Rebecca Saxe, "How We Read Other People's Minds" (Saxe, 2009). Many of Saxe's papers are to be found on her Saxelab website, http://saxelab.mit.edu/.

Chapter 5: The Natural History of Historians

Robert Axelrod's *Evolution of Cooperation* (Axelrod, 1984) and its sequel, *The Complexity of Cooperation* (Axelrod, 1997) were landmarks in the study of how a Darwinian process harnesses fitness maximization to produce altruistic behavior. Robert Frank's *Passions within Reason* (Frank, 1988) brought into the discussion the role of emotions in favoring prosocial behavior.

Then a team of social psychologists, experimental economists, evolutionary anthropologists, and game theorists undertook a cross-cultural empirical study that strongly substantiated their insights: *Foundations of Human Sociality: Economic Experiments and Ethnographic Evidence from Fifteen Small-Scale Societies* (Henrich et al., 2004) edited by Joseph Henrich, Robert Boyd, Samuel Bowles, Colin Camerer, Ernst Fehr, and Herbert Gintis. Each of the editors of this collected volume has written important works on the subject of human sociality both before and after the publication of their team's cross-cultural research.

How such prosocial inclinations and dispositions could have started out as learned and ended up as innate and triggered by a slight stimulus in early infancy is a matter of speculation. But the Baldwin effect is one widely discussed possibility. A wide-ranging but tendentious introduction to the methods and findings about innate and nearly innate cognition, emotion, and sensory processing is David Buss's *Evolutionary Psychology: The New Science of the Mind* (Buss, 2016), which has gone through numerous editions, updatings, and revisions.

The most influential experimental and observational studies that have shed light on human cognitive evolution and how hominins diverged from the other primates have been undertaken by Michael Tomasello, as expounded in his *A Natural History of Human Thinking* (Tomasello, 2014) and *A Natural History of Human Morality* (Tomasello, 2016) and by Sarah Hrdy in her *Mothers and Others: The Evolutionary Origins of Mutual Understanding* (Hrdy, 2009). Tomasello's and Hrdy's work has converged over the years; their important collaborations have provided the best available account of how the theory of mind could have emerged from the mind-reading ability common to other mammals.

Chapter 6: What Exactly Was the Kaiser Thinking?

The indeterminacy of psychological states by behavior has been a fixed and accepted conclusion among philosophers of science since the work of Willard Van Orman Quine, especially his *Word and Object* (Quine, 1960). Donald Davidson made the problems explicit in many articles, some collected in *Essays on Action and Events* (Davidson, 1980). The limitations this conclusion imposes on psychology as a science are most starkly posed in many of the works of Daniel Dennett, especially *Content and Consciousness* (Dennett, 1969), *Brainstorms* (Dennett, 1981), and *The Intentional Stance* (Dennett, 1987). The problem of indeterminacy is, in part, what unraveled Skinnerian behavioral psychology and led to the emergence of cognitive science.

Among philosophers of mind, John Searle is the most vocal defender of the claim that conscious awareness is a reliable guide to the specific content of at least some of our beliefs and desires. This is a view Searle has held over a long span of works, starting with his *Intentionality: An Essay in the Philosophy of Mind* (Searle, 1983).

Bernard Baars has been developing an influential psychological theory of the nature and function of consciousness for decades. Together with Nicole M. Gage, he has written a text that integrates this work with the broader subject. *Cognition, Brain and Consciousness: An Introduction to Cognitive Neuroscience* (Baars and Gage, 2010). Jesse Prinz's *The Conscious Brain* is a great philosopher's argument for consciousness as attention.

Chapter 7: Can Neuroscience Tell Us What Talleyrand Meant?

The best way to begin to get a grip on the problem that representation poses for psychology is to read the *Stanford Encyclopedia* article, "Mental Representation" at https://plato.stanford.edu/entries/mental-representation/. Alas, it tells us almost nothing about how the brain represents.

The story of how the medical mistreatment of a single patient, H.M., led to the entire industry of hippocampal research is told in typically narrative mode by Luke Dittrich in *Patient H.M.: A Story of Memory, Madness and Family Secrets* (Dittrich, 2016). Suzanne Corkin's *Permanent Present Tense: The Unforgettable Life of the Amnesic Patient, H.M.* reports on much of the science studying H.M. produced.

The work of Eric Kandel, John O'Keefe, and May-Britt and Edvard Moser is well described in their Nobel Lectures, all of which are available online as PDF files (Kandel, 2000; O'Keefe, 2014; M.-B. Moser, 2014; and E. I. Moser, 2014). There are also videos of their lectures available online: Kandel's at https://www.nobelprize.org/nobel_prizes/medicine/laureates/2000/kandel-lecture.html; O'Keefe's at https://www.nobelprize.org/mediaplayer/?id =2413/; May-Britt Moser's at https://www.nobelprize.org/nobel_prizes /medicine/laureates/2014/may-britt-moser-lecture.html; and Edvard Moser's at https://www.nobelprize.org/mediaplayer/index.php?id=2426/.

Chapter 8: Talleyrand's Betrayal: The Inside Story

The work of O'Keefe and the Mosers triggered an avalanche of further research on the details of how place and grid cells function. More important, it provided tools and motivation to explore the way the same brain regions store other details about the environment and deploy them in behavior. Among the most interesting of these research programs is the work of Loren Frank, whose lab's website, http://www.cin.ucsf.edu/HTML /Loren_Frank.html, includes a link to "Lab Discoveries." Although the papers listed there are almost all technical lab reports, Frank has also made highly accessible videos; "What Is Memory?" which uses animation to make its points, is particularly worth watching, https://www.youtube.com/ watch?v=V1KOiOVUydo/.

Well before O'Keefe and the Mosers made the hippocampus famous, it was the subject of the long research career of the late Howard Eichenbaum, whose work is available on his lab's website, https://www.bu.edu/cogneuro/.

Chapter 9: *Jeopardy!* Answer: "It Shows the Theory of Mind to Be Completely Wrong"

The comparison of mind to software and brain to hardware goes back to a famous 1950 paper by Alan Turing, "Computing Machinery and Intelligence" (Turing, 1950), widely reprinted and available online. Philosophers and others have vigorously disputed whether the mind is or is not a matter of the brain running a program since soon after Turing wrote his paper. As usual, a great place to start exploring the issues is "The Computational Theory of Mind" by Michael Rescorla in the *Stanford Encyclopedia* (Rescorla, 2015). Perhaps the most famous argument against the theory that cognition is largely a matter of implementing a program was advanced in John Searle's paper "Minds, Brains, and Programs" (Searle, 1980).

A great of deal of subsequent science fiction in print, film, and video has proceeded on the assumption that Searle's conclusions are wrong (without, however, addressing his arguments). Greg Egan's science fiction novel *Permutation City* (Egan, 1984) is an influential example. "San Junipero," an episode of the science fiction anthology TV series *Blackmirror* is another example of artists taking the Turing theory seriously.

The IBM website is one place to learn more about Deep Blue and Watson. Both, of course, are the subjects of narrative story–based science writing. Typical of the genre is Stephen Baker's *Final Jeopardy: The Story of Watson, the Computer That Will Transform Our World* (Baker, 2012).

Philosophy has been wrestling with the problem of how cognition could be physical since Descartes in the seventeenth century. Two wonderful anthologies between them bring most of the important contributions together, William Lycan's *Mind and Cognition* (Lycan, 1990) and John Heil's *Philosophy of Mind* (Heil, 2004). Alas, neither incudes Leibniz's illuminating thought experiment (section 17 of his essay "Monadology," cited in chapter 9).

Chapter 10: The Future of an Illusion

The project of applying Darwinian natural selection to understanding the appearance of purpose in human thought starts with Dan Dennett's *Content and Consciousness* (Dennett, 1969) and Larry Wright's *Teleological Explanations* (Wright, 1976). It made great strides in Ruth Millikan's *Language, Thought and Other Biological Categories* (Millikan, 1984) and reached an apogee in Karen Neander's *A Mark of the Mental* (Neander, 2017). The place to start is Neander's article "Teleological Theories of Mental Content" in the *Stanford Encyclopedia of Philosophy* (Neander, 2004; rev. 2012). Jerry Fodor's *A Theory of Content and Other Essays* (Fodor, 1990) identifies the obstacles to teleosemantics as an account of mental representation.

Patricia Churchland's *Neurophilosophy: Toward a Unified Science of the Mind/Brain* (Churchland, 1986) pretty much originated the insight that the theory of mind was a dead end, a wrong turn, an obstacle to the development of a science of psychology.

In 1866, the Linguistic Society of Paris formally banned any debate regarding the origin of language, the problem being both too vexed and too daunting. Evolutionary psychology and Noam Chomsky's work in linguistics have at least clarified the issues. The best discussion (and a great example of science without stories) can be found in Steven Pinker's *Language Instinct* (Pinker, 1994) and *The Stuff of Thought* (Pinker, 2007).

Chapter 11: Henry Kissinger Mind Reads His Way through the Congress of Vienna

Kissinger has been riding the hobbyhorse of the Congress of Vienna for almost sixty years, all the way from his 1956 article in *World Politics* (Kissinger, 1956) to his 2014 book, *World Order* (Kissinger, 2014). Experience doesn't seem to have had much effect on his attachment to the theory of mind or on his applying it to theories of international politics. The steady sales of reprint editions of Christopher Hitchens's *The Trial of Henry Kissinger* (Hitchens, 2012 [2001]) on both sides of the Atlantic ever since the book was first published in 2001 testify to an insatiable demand for a narrative explanation of Kissinger's interventions in American foreign policy from Vietnam onward.

Chapter 12: *Guns, Germs, Steel*—and All That

Like so many other dimensions of intellectual life, history is being colonized by the biological sciences, substituting Darwinian approaches to explanation for ones based on narratives. Economic history was perhaps the earliest manifestation of the trend to become publicly controversial. In the 1970s, Robert Fogel and Stanley Engerman's *Time on the Cross* (Fogel and Engerman, 1974) sparked a raging debate about the nature of American slavery and its role in American history. Peter Turchin has applied a variety of demographic and ecological theories to explain human history in a series of works on several time scales from millennia in *War and Peace and War* (Turchin, 2006) to decades of American history in *Ages of Discord* (Turchin, 2016). Another arresting example of this sort of history is Gregory Clark's *A Farewell to Alms* (Clark, 2007), which provides a powerful nonnarrative explanation for the emergence of the Industrial Revolution first in Britain.

Chapter 13: *The Gulag Archipelago* and the Uses of History

Every national movement, religious tradition, regional culture, ethnic group, political party, and sports team has its own, often inflammatory narrative history. Some of them are entertaining. Few of them have any wisdom—and none, any knowledge—to impart.

References

Axelrod, R. (1984). *Evolution of Cooperation*. New York: Basic Books.

Axelrod, R. (1997). *The Complexity of Cooperation: Agent-Based Models of Competition and Collaboration*. Princeton: Princeton University Press.

Azizi, A. H., Wiskott, L., & Cheng, S. (2013). A Computational Model for Preplay in the Hippocampus. *Frontiers in Computational Neuroscience, 7*, 161. doi:10.3389/fncom.2013.00161.

Baars, B. J. (1997). *In the Theater of Consciousness: The Workspace of the Mind*. New York: Oxford University Press.

Baars, B. J. (2002). The Conscious Access Hypothesis: Origins and Recent Evidence. *Trends in Cognitive Sciences, 6*(1), 47–52.

Baars, B. J., & Gage, N. M. (2010). *Cognition, Brain, and Consciousness: An Introduction to Cognitive Neuroscience*. London: Elsevier Academic Press.

Bae, B.-I., Jayaraman, D., & Walsh, C. A. (2015). Genetic Changes Shaping the Human Brain. *Developmental Cell, 32*(4), 423–434. doi:10.1016/j.devcel.2015.01.035.

Bailey, C. H., Bartsch, D., & Kandel, E. R. (1996). Toward a Molecular Definition of Long-Term Memory Storage. *Proceedings of the National Academy of Sciences of the United States of America, 93*(24), 13445–13452.

Baker, S. (2012). *Final Jeopardy: The Story of Watson, the Computer That Will Transform Our World*. Boston: Mariner Books.

Baldwin, D. A. (1991). Infants' Contribution to the Achievement of Joint References. *Child Development, 62*(5), 874–890.

Baldwin, D. A., Markman, E. M., & Melartin, R. L. (1993). Infants' Ability to Draw Inferences about Nonobvious Object Properties: Evidence from Exploratory Play. *Child Development, 64*(3), 711–723.

Baron-Cohen, S. (1995). *Mindblindness: An Essay on Autism and the Mind*. Cambridge, MA: MIT Press.

Baron-Cohen, S., Leslie, A. M., & Frith, U. (1985). Does the Autistic Child Have a "Theory of Mind"? *Cognition, 21*(1), 37–46. doi:10.1016/0010-0277(85)90022-8.

Bateson, M., Nettle, D., & Roberts, G. (2006). Cues of Being Watched Enhance Cooperation in a Real-World Setting. *Biology Letters, 2*(3), 412–414.

Beard, M. (2015). *SPQR: A History of Ancient Rome*. London: Profile Books.

Ben-Yakov, A., Dudai, Y., & Mayford, M. R. (2015). Memory Retrieval in Mice and Men. *Cold Spring Harbor Perspectives in Biology, 7*, a021790. doi:10.1101/cshperspect. a021790.

Bloom, P. (2001). Précis of *How Children Learn the Meanings of Words. Behavioral and Brain Sciences, 24*(6), 1095–1103. http://groups.psych.northwestern.edu/waxman /wordextension_CommPBloom.pdf.

Boyer, P. (2008). Being Human: Religion: Bound to Believe? *Nature, 455*(7216), 1038–1039.

Boyer, P., & Bergstrom, B. (2008). Evolutionary Perspectives on Religion. *Annual Review of Anthropology, 37*, 111–130.

Brandler, W. M., & Sebat, J. (2015). From de novo Mutations to Personalized Therapeutic Interventions in Autism. *Annual Review of Medicine, 66*(1), 487–507. doi:10.1146/annurev-med-091113-024550.

Brinton, C. (1963). *Lives of Talleyrand*. New York: Norton. (1936).

Bryson, B. (2003). *A Short History of Nearly Everything*. New York: Broadway Books.

Burkhart, J. M., Hrdy, S. B., & van Schaik, A. N. D. C. P. (2009). Cooperative Breeding and Human Cognitive Evolution. *Evolutionary Anthropology, 18*, 175–186.

Buss, D. M. (2016). *Evolutionary Psychology: The New Science of the Mind*. London: Routledge.

Carroll, S. (2010). *From Eternity to Here: The Quest for the Ultimate Theory of time*. New York: Dutton.

Carruthers, P. (2015). *The Opacity of Mind: An Integrative Theory of Self-knowledge*. Oxford: Oxford University Press.

Churchill, W. (1986). *The Second World War* (6 Vols.). New York: Houghton Mifflin/ Mariner Books. (1948).

Churchland, P. S. (1986). *Neurophilosophy: Toward a Unified Science of the Mind/Brain*. Cambridge, MA: MIT Press.

Clark, G. (2007). *A Farewell to Alms: A Brief Economic History of the World.* Princeton: Princeton University Press.

Cloud, D. (2014). *Domestication of Language: Cultural Evolution and the Uniqueness of the Human Animal.* New York, NY: Columbia University Press.

Cooper, D. (2001). *Talleyrand.* New York: Grove Press. (1932).

Darwin, C. (1871). *The Descent of Man, and Selection in Relation to Sex* (2 Vols.). London: John Murray.

Darwin, C. (1872). *On the Origin of Species by Means of Natural Selection, or the Preservation of Favoured Races in the Struggle for Life* (6th ed.). London: John Murray. (1859).

Davidson, D. (1980). *Essays on Action and Events.* New York Oxford University Press.

Dawkins, R. (1976). *The Selfish Gene.* Oxford: Oxford University Press.

Dehaene, S., & Naccache, L. (2001). Towards a Cognitive Neuroscience of Consciousness: Basic Evidence and a Workspace Framework. *Cognition, 79*(1–2), 1–37.

Dennett, D. C. (1969). *Content and Consciousness.* London: Routledge & Kegan Paul.

Dennett, D. C. (1978). Beliefs about Beliefs. *Behavioral and Brain Sciences, 1*(4), 568–570.

Dennett, D. C. (1981). *Brainstorms: Philosophical Essays Of Mind Psychology.* Cambridge, MA: MIT Press.

Dennett, D. C. (1987). *The Intentional Stance.* Cambridge, MA: MIT Press.

Diamond, J. (1997). *Guns, Germs, and Steel: The Fates of Human Societies.* New York: Norton.

Diamond, J. (n.d.) "Geographic Determinism." http://www.jareddiamond.org/Jared_Diamond/Geographic_determinism.html

Dittrich, L. (2016). *Patient H.M.: A Story of Memory, Madness, and Family secrets.* New York: Random House.

Dragoi, G. (2013). Internal Operations in the Hippocampus: Single Cell and Ensemble Temporal Coding. *Frontiers in Systems Neuroscience, 7,* 46. doi:10.3389/fnsys.2013.00046.

Edgeworth, F. Y. (1881). *Mathematical Psychics: An Essay on the Application of Mathematics to the Moral Sciences.* London: Kegan Paul.

Egan, G. (1994). *Permutation City.* New York: HarperPrism.

Eichenbaum, H. (2013). Hippocampus: Remembering the Choices. *Neuron, 77*(6), 999–1001. doi:10.1016/j.neuron.2013.02.034.

Eisenhower, J. S. D. (1969). *The Bitter Woods.* New York: Putnam.

Everett, D. L. (2005). Cultural Constraints on Grammar and Cognition in Pirahã: Another Look at the Design Features of Human Language. *Current Anthropology, 46*(4), 621–646. doi:10.1086/431525.

Falk, D., Zollikofer, C. P. E., Morimoto, N., & Ponce de León, M. S. (2012). Metopic Suture of Taung (*Australopithecus africanus*) and its Implications for Hominin Brain Evolution. *Proceedings of the National Academy of Sciences of the United States of America, 109*(22), 8467–8470. doi:10.1073/pnas.1119752109

Faulkner, W. (1951). *Requiem for a Nun.* New York: Random House.

Fodor, J. (1975). *The Language of Thought.* New York: Thomas Crowell.

Fodor, J. (1983). *The Modularity of Mind: An Essay on Faculty Psychology.* Cambridge, MA: MIT Press.

Fodor, J. A. (1987). *Psychosemantics: The Problem of Meaning in the Philosophy of Mind.* Cambridge, MA: MIT Press.

Fodor, J. (1990). *A Theory of Content and Other Essays.* Cambridge, MA: MIT Press.

Fogel, R. W., & Engerman, S. L. (1974). *Time on the Cross: The Economics of American Negro Slavery.* Boston: Little, Brown.

Frank, R. (1988). *Passions within Reason: The Strategic Role of the Emotions.* New York: Norton.

Freud, S. (1927) (1989). In P. Gay (Ed.), *The Future of an Illusion. Standard edition* (J. Strachey, Trans.). New York: Norton.

Friedman, M. (1953). *Essays on Positive Economics.* Chicago, IL: University of Chicago Press.

Gibbon, E. (1776/1789). *The History of the Decline and Fall of the Roman Empire* (6 vols.) London: Stahan and Cadell.

Gigerenzer, G., & Gaissmeier, G. (2011). Heuristic Decision Making. *Annual Review of Psychology, 62,* 451–482. doi:10.1146/annurev-psych-120709-145346.

Girardeau, G., & Zugaro, M. (2011). Hippocampal Ripples and Memory Consolidation. *Current Opinion in Neurobiology, 21*(3), 452–459. doi:10.1016/j.conb.2011.02.005.

Glimcher, P. W., & Fehr, E. (2012). *Neuroeconomics: Decision Making and the Brain* (2nd ed.). New York, NY: Academic Press.

Goldman, A. I. (2006). *Simulating Minds: The Philosophy, Psychology, and Neuroscience of Mindreading.* Oxford: Oxford University Press.

Gopnik, A. (2000). Explanation as Orgasm and the Drive for Causal Understanding: The Evolution, Function and Phenomenology of the Theory-formation System. In F. Keil & R. Wilson (Eds.), *Cognition and Explanation* (pp. 299–323). Cambridge, MA: MIT Press.

Gredebäck, G., & Melinder, A. (2011). Teleological Reasoning in 4-month-old Infants: Pupil Dilations and Contextual Constraints. *PLoS One, 6*(10), e26487.

Greene, B. (1999). *The Elegant Universe: Superstrings, Hidden Dimensions, and the Quest for the Ultimate Theory*. New York: Norton.

Greene, B. (2004). *The Fabric of the Cosmos: Space, Time, and the Texture of Reality*. New York: Knopf.

Grice, P. (1991). *Studies in the Way of Words*. Cambridge, MA: Harvard University Press.

Gweon, H., & Saxe, R. (2013). Developmental Cognitive Neuroscience of Theory of Mind. In J. Rubenstein & P. Rakic (Eds.), *Neural Circuit Development and Function in the Brain: Comprehensive Developmental Neuroscience* (pp. 367–377). San Diego, CA: Academic Press. http://saxelab.mit.edu/resources/papers/Gweon_Saxe_2013.pdf.

Hartley, T., Lever, C., Burgess, N., & O'Keefe, J. (2014). Space in the Brain: How the Hippocampal Formation Supports Spatial Cognition. *Philosophical Transactions of the Royal Society of London. Series B, Biological Sciences, 369*, 20120510. doi:10.1098/rstb.2012.0510.

Haugeland, J. (1990). The Intentionality All-Stars. In J. E. Tomberlin (Ed.), *Philosophical Perspectives (Vol. 4) Action Theory and Philosophical of Mind* (pp. 383–437). Atascadero, CA: Ridgeview. http://people.exeter.ac.uk/sp344/int%20all%20stars.pdf.

Hawking, S. (1988). *A Brief History of Time*. New York: Bantam Books.

Heil, J. (Ed.). (2004). *Philosophy of Mind: A Guide and Anthology*. Oxford: Oxford University Press.

Henke, K. (2010). A Model for Memory Systems Based on Processing Modes Rather Than Consciousness. *Nature Reviews. Neuroscience, 11*(7), 523–532. http://www.nature.com/nrn/journal /v11/n7/abs/nrn2850.html.

Henrich, J., Boyd, R., Bowles, S., Camerer, C., Fehr, E., & Gintis, H. (Eds.). (2004). *Foundations of Human Sociality: Economic Experiments and Ethnographic Evidence from Fifteen Small-Scale Societies*. Oxford: Oxford University Press.

Hitchens, C. (2012). *The Trial of Henry Kissinger*. New York: Twelve. (2001).

Homer. (1998). *The Iliad* (R. Fagles, Trans.). New York: Penguin Books.

Horne, A. (1965). *The Fall of Paris: The Siege and the Commune, 1870–71*. London: Macmillan.

Horne, A. (1969). *To Lose a Battle: France, 1940*. London: Macmillan.

Hortensius, R., de Gelder, B., & Schutter, D. J. L. G. (2016). When Anger Dominates the Mind: Increased Motor Corticospinal Excitability in the Face of Threat. *Psychophysiology, 53*, 1307–1316. doi:10.1111/psyp.12685.

Hrdy, S. (2009). *Mothers and Others: The Revolutionary Origins of Mutual Understanding*. Cambridge, MA: Harvard University Press.

Hublin, J.-J., Ben-Ncer, A., Bailey, S. E., Freidline, S. E., Neubauer, S., Skinner, M. M., et al. (2017). New Fossils from Jebel Irhoud, Morocco and the pan-African Origin of *Homo sapiens*. *Nature, 546*, 289–292. doi:10.1038/nature22336.

Hume, D. (1985). In E. C. Mossner (Ed.), *A Treatise of Human Nature*. Harmondsworth: Penguin Books. [1739–1740]

Hume, D. (2007 [1748]). Of the idea of necessary connexion. In P. Millican (Ed.), *An Enquiry Concerning Human Understanding* (sec. 7: pp. 42–46). Oxford: Oxford University Press.

Isaacson, W. (2011). *Steve Jobs: The Exclusive Biography*. New York: Simon & Schuster.

Jack, A., Connelly, J. J., & Morris, J. P. (2012). DNA Methylation of the Oxytocin Receptor Gene Predicts Neural Response to Ambiguous Social Stimuli. *Frontiers in Human Neuroscience, 6*, 280.

Jadhav, S. P., & Frank, L. M. (2014). Memory Replay in the Hippocampus. In D. Derdikman & J. J. Knierim (Eds.), *Space, Time and Memory in the Hippocampal Formation* (pp. 351–371). Vienna, Austria: Springer; 10.1007/978-3-7091-1292-2_13.

Kahneman, D., & Thaler, R. H. (2006). Anomalies: Utility Maximization and Experienced Utility. *Journal of Economic Perspectives, 20*(1), 221–234.

Kalbe, E., Schlegel, M., Sack, A. T., Nowak, D. A., Dafotakos, M., Bangard, C., et al. (2010). Dissociating Cognitive from Affective Theory of Mind: A TMS Study. *Cortex, 46*(6), 769–780. doi:10.1016/j.cortex.2009.07.010.

Kandel, E. R. (2000). "The Molecular Biology of Memory Storage: A Dialog between Genes and Synapses." Nobel Lecture. December 8. https://www.nobelprize.org/nobel_prizes/medicine/laureates/2000/kandel-lecture.pdf.

Kay, K., Sosa, M., Chung, J. E., Karlsson, M. P., Larkin, M. C., & Frank, L. M. (2016). A Hippocampal Network for Spatial Coding during Immobility and Sleep. *Nature, 531*(7593), 185–190. doi:10.1038/nature17144.

Kershaw, I. (2007). *Fateful Choices*. New York: Penguin Press.

Kissinger, H. (1956). The Congress of Vienna: A Reappraisal. *World Politics, 8*(2), 264–280.

Kissinger, H. (1957). *A World Restored: Metternich, Castlereagh, and the Problems Of Peace, 1812–22*. Boston: Houghton Mifflin.

Kissinger, H. (2014). *World Order*. New York: Penguin Press.

Kitamura, T., Sun, C., Martin, J., Kitch, L. J., Schnitzer, M. J., & Tonegawa, S. (2015). Entorhinal Cortical Ocean Cells Encode Specific Contexts and Drive Context-specific Fear Memory. *Neuron, 87*(6), 1317–1331. doi:10.1016/j.neuron.2015.08.036.

Kitcher, P. (1992). *Freud's Dream: A Complete Interdisciplinary Science of Mind*. Cambridge, MA: MIT Press.

Krupenye, C., Kano, F., Hirata, S., Call, J., & Tomasello, M. (2016). Cognition: Great Apes Anticipate That Other Individuals Will Act According to False Beliefs. *Science, 354*(6308), 110–114. doi:10.1126/science.aaf8110.

Kubie, J. L., & Fox, S. E. (2015). Do the Special Frequencies of Grid Cells Mold the Firing Fields of Place Cells? *Proceedings of the National Academy of Sciences of the United States of America, 118*(13), 3860–3861. http://www.pnas.org/content/112/13/3860.

Leibniz, G. W. (1981). Preface. In P. Remnant & J. Bennett Ed. & Trans. *New Essays on Human Understanding* (pp. 44–68). Cambridge: Cambridge University Press. (1765).

Leibniz, G. W. (1965). Monadology. In P. Schrecker & A. M. Schrecker Ed. & Trans. *Monadology and Other Essays* (pp. 148–165). Indianapolis: Bobbs-Merrill. (1714).

Libet, B. (1985). Unconscious Cerebral Initiative and the Role of Conscious Will in Voluntary Action. *Behavioral and Brain Sciences, 8*(4), 529–566.

Lowenstein, R. (2015). *America's Bank: The Epic Struggle to Create the Federal Reserve*. New York: Penguin Press.

Lurz, R. W. (2011). *Mindreading Animals: The Debate over What Animals Know About Other Minds*. Cambridge, MA: MIT Press.

Lycan, W. G. (Ed.). (1990). *Mind and Cognition*. London: Blackwell.

Mackie, G. (1996). Ending Foot Binding and Infibulation: A Conventional Account. *American Sociological Review, 61*, 999–1017.

Malle, B. F. (2002). "The Relation between Language and Theory of Mind in Development and Evolution," in T. Givón & B. F. Malle (Eds.), *The Evolution of Language out of Pre-language* (pp. 265–284). Amsterdam: Benjamins.

Manns, J. R., & Eichenbaum, H. (2006). Evolution of Declarative Memory. *Hippocampus, 16*(3), 795–808.

Marouli, E., Graff, G., Medina-Gomez, C., Sin Lo, K., Wood, A. R., Kjaer, T. R., et al. (2017). Rare and Low-frequency Coding Variants Alter Human Adult Height. *Nature, 542*, 186–190. doi:10.1038/nature21039.

McPherson, J. (1988). *Battle Cry Of Freedom: The Civil War Era*. New York: Oxford University Press.

McPherson, J. (2009). *This Mighty Scourge: Perspectives on the Civil War*. Oxford: Oxford University Press.

Meltzoff, S., Gopnik, A., & Kuhl, P. K. (1999). *The scientist in the Crib: What Early Learning Tells Us about the Mind*. New York: William Morrow.

Millikan, R. (1984). *Language, Thought, and Other Biological Categories*. Cambridge, MA: MIT Press.

Moser, E. I. (2014) "Grid Cells and the Entorhinal Map of Space." Nobel Lecture. December 7. https://www.nobelprize.org/nobel_prizes/medicine/laureates/2014/edvard -moser-lecture-slides.pdf.

Moser, E. I., Roudi, Y., Witter, M. P., Kemtros, C., Bonhoeffer, T., & Moser, M.-B. (2014). Grid Cells and Cortical Representation. *Nature Reviews. Neuroscience*, *15*, 466–481. doi:10.1038/nrn376.

Moser, M.-B. (2014). "Grid Cells, Place Cells, and Memory." Nobel Lecture. December 7. https://www.nobelprize.org/nobel_prizes/medicine/laureates/2014/may-britt -moser-lecture-slides.pdf.

Nagel, T. (1974). What Is It Like to Be a Bat? *Philosophical Review*, *83*(4), 435–450.

Neander, K. (2004; rev. 2012). "Teleological Theories of Mental Content." In *Stanford Encyclopedia of Philosophy*. https://plato.stanford.edu/entries/content-teleological/.

Neander, K. (2017). *A Mark of the Mind: In defense of informational telesemantics*. Cambridge, MA: MIT Press.

Nguyen, P. V., & Kandel, E. R. (1996). A Macromolecular Synthesis-Dependent Late Phase of Long-term Potentiation Requiring cAMP in the Medial Perforant Pathway of Rat Hippocampal Slices. *Journal of Neuroscience*, *16*(10), 3183–3198.

Nichols, S., Stich, S., Leslie, A., & Klein, D. (1996). Varieties of Off-line Simulation. In P. Carruthers & P. Smith (Eds.), *Theories of Theories Of Mind* (pp. 39–74). Cambridge: Cambridge University Press; http://cogprints.org/376/1/sim3.html.</edb.

O'Connell, G., Hsu, C.-T., Christakou, A., and Chakrabarti, B. (2018). Thinking about Others and the Future: Neural Correlates of Perspective Taking Relate to Preferences for Delayed Rewards. *Cognitive, Affective & Behavioral Neuroscience*, *18*(1), 35–42. doi:10.3758/s13415-017-0550-8

Oesch, N., and Dunbar, R. I. M. (2016). The Emergence of Recursion in Human Language: Mentalizing Predicts Recursive Syntax Task Performance. *Journal of Neurolinguistics*, *43*, 95–106.

Öhman, A., & Mineka, S. (2001). Fears, Phobias, and Preparedness. Toward an Evolved Module of Fear and Fear Learning. *Psychological Review, 108*(3), 483–522.

O'Keefe, J. (2014). "Spatial Cells in the Hippocampal Formation." Nobel Lecture. December 7. https://www.nobelprize.org/nobel_prizes/medicine/laureates/2014/okeefe-lecture-slides.pdf.

O'Keefe, J., & Dostrovsky, J. (1971). The Hippocampus as a Spatial Map: Preliminary Evidence from Unit Activity in the Freely-moving Rat. *Brain Research, 34*(1), 171–175.

Omer, D. B., Maimon, S. R., Las, L., & Ulanovsky, N. (2018). Social Place-cells in the Bat Hippocampus. *Science, 359*, 218–224. doi:10.1126/science.aao3474.

Paley, W. (1802). *Natural Theology, or Evidences of the Existence and Attributes of the Deity.* London: R. Faulder.

Piketty, T. (2014). *Capital in the Twenty-first Century* (A. Goldhammer, Trans.). Cambridge, MA: Harvard University Press.

Pinker, S. (1994). *Language Instinct.* New York: Harper Perennial.

Pinker, S. (1997). *How the Mind Works.* New York: Norton.

Pinker, S. (2007). *The Stuff of Thought: Language as a Window into Human Nature.* New York: Viking.

Poulin-Dubois, D., Sodian, B., Metz, U., Tilden, J., & Schoeppner, B. (2007). Out of Sight Is not Out of Mind: Developmental Changes in Infants' Understanding of Visual Perception during the Second Year. *Journal of Cognition and Development, 8*(4), 401–425. doi:10.1080/15248370701612951.

Powell, A., Shennan, S., & Thomas, M. G. (2009). Late Pleistocene Demography and the Appearance of Modern Human Behavior. *Science, 324*(5932), 1298–1301. doi:10.1126/science.1170165.

Prinz, J. J. (2012). *The Conscious Brain: How Attention Engenders Experience.* New York, NY: Oxford University Press.

Quine, W. V. O. (1960). *Word and Object.* Cambridge, MA: MIT Press.

Raichle, M. E., MacLeod, A. M., Snyder, A. Z., Powers, W. J., Gusnard, D. A., & Shulman, G. L. (2001). A Default Mode of Brain Function. *Proceedings of the National Academy of Sciences of the United States of America, 98*(2), 676–682. doi:10.1073/pnas.98.2.676.

Ravenscroft, J. (1997; rev. 2016). "Folk Psychology as a Theory." In *Stanford Encyclopedia of Philosophy.* https://plato.stanford.edu/entries/folkpsych-theory/

Rescorla, M. (2015). "The Computational Theory of the Mind." In *Stanford Encyclopedia of Philosophy.* https://plato.stanford.edu/entries/computional-mind/

Rieff, D. (2016). *In Praise of Forgetting: Historic Memory and its Ironies*. New Haven: Yale University Press.

Robbins, L. (1932). *An Essay on the Nature and Significance of Economic Science*. London, United Kingdom: Macmillan.

Rood, T. (1998). *Thucydides:Narrative and Explanation*. Oxford: Oxford University Press.

Roumis, D. K., & Frank, L. M. (2015). Hippocampal Sharp Wave Ripples in Waking and Sleeping States. *Current Opinion in Neurobiology, 35*, 6–12. doi:10.1016/j. conb.2015.05.001.

Rubin, R. (2015). "'America's Bank,' by Roger Lowenstein." *New York Times*, October 19.

Santayana, G. (1905). *Reason in Common Sense*. Vol. *1* of The Life of Reason; or, the Phases of Human Progress. New York: Scribner.

Saxe, R. (2009). "How We Read Other People's Minds." TED talk. July. https://www .ted.com/talks/rebecca_saxe_how_brains_make_moral_judgments/

Saxe, R., Carey, S., & Kanwisher, N. (2004). Understanding Other Minds: Linking Developmental Psychology and Functional Neuroimaging. *Annual Review of Psychology, 55*, 87–124.

Schlender, B., & Tetzeli, R. (2015). *Becoming Steve Jobs: The Evolution of a Reckless Upstart into a Visionary Leader*. New York: Crown Business.

Schuwerk, T., Langguth, B., & Sommer, M. (2014). Modulating Functional and Dysfunctional Mentalizing by Transcranial Magnetic Stimulation. *Frontiers in Psychology, 5*, 1309. http://www.ncbi.nlm.nih.gov/pmc/articles/PMC4235411/.

Searle, J. R. (1980). Minds, Brains, and Programs. *Behavioral and Brain Sciences, 3*(3), 417–424. doi:10.1017/S0140525X00005756

Searle, J. R. (1983). *Intentionality: An Essay in the Philosophy of Mind*. Cambridge, MA: Harvard University Press.

Simard, I., Luck, D., Mottron, L., Zeffiro, T. A., & Soulières, I. (2015). Autistic Fluid Intelligence: Increased Reliance on Visual Functional Connectivity with Diminished Modulation of Coupling by Task Difficulty. *NeuroImage. Clinical, 9*, 467–478. doi:10.1016/j.nicl.2015.09.007.

Skyrms, B. (1996). *Evolution of the Social Contract*. Cambridge, United Kingdom: Cambridge University Press.

Skyrms, B. (2004). *The Stag Hunt and Evolution of Social Structure*. Cambridge, United Kingdom: Cambridge University Press.

Solzhenitsyn, A. I. (1973–1978). *The Gulag Archipelago, 1918–1956: An Experiment in Literary Investigation* (3 vols.). T. P. Whitney (Trans.). New York: Harper & Row.

Southey, R. ([1813] [1917]). *The Life of Nelson.* London: Nelson.

Sterelny, K. (2003). *Thought in a Hostile World: The Evolution of Cognition.* Malden, MA: Blackwell.

Sterelny, K. (2012). *The Evolved Apprentice: How Evolution Made Humans Unique.* Cambridge, MA: MIT Press.

Stich, S. (1983). *From Folk Psychology to Cognitive Science.* Cambridge, MA: MIT Press.

Stringer, C. (2013). *Lone Survivors: How We Came to be the Only Humans on Earth.* London: St. Martin's Press.

Teaford, M. F., and Ungar, P. S. (2000). Diet and the Evolution of the Earliest Human Ancestors. *Proceedings of the National Academy of Sciences of the United States of America, 97*(25), 13506–13511. doi:10.1073/pnas.260368897

Tolman, E. C. (1948). Cognitive Maps in Rats and Men. *Psychological Review, 53*(4), 189–208.

Tomasello, M. (2014). *A Natural History of Human Thinking.* Cambridge, MA: Harvard University Press.

Tomasello, M. (2016). *A Natural History of Human Morality.* Cambridge, MA: Harvard University Press.

Tuchman, B. W. (1962). *The Guns of August.* New York: Macmillan.

Turchin, P. (2006). *War and Peace and War: The Rise and Fall of Empires.* New York: Plume.

Turchin, P. (2016). *Ages of Discord: A Structural-demographic Analysis of American History.* Chaplin, CT: Beresta Books.

Turing, A. M. (1950). Computing Machinery and Intelligence. *Mind, 59*(236), 433–460. doi:10.1093/mind/LIX.236.433.

Watson, J. D. (1968). *The Double Helix: A Personal Account of the Discovery of the Structure of DNA.* New York: Atheneum.

Weinberg, G. L. (2012). World War II. In R. Chickering, D. Showalter, and H. van de Ven (Eds.), *War and the Modern World.* Vol. 4 of The Cambridge History of War. Cambridge: Cambridge University Press.

Winchester, S. (2003). *Krakatoa: The Day the World Exploded, August 27, 1883.* New York: HarperCollins.

Wright, L. (1976). *Teleological Explanations: An Etiological Analysis of Goals and Functions. Berkely.* University of California Press.

Yu, J. Y., & Frank, L. M. (2015). Hippocampal-cortical Interaction in Decision Making. *Neurobiology of Learning and Memory, 117*(01501), 34–41. https://www.ncbi.nlm.nih.gov/pubmed/24530374.

Yun, K., Chung, D., Jang, B., Kim, J. H., & Jeong, J. (2011). Mathematically Gifted Adolescents Have Deficiencies in Social Valuation and Mentalization. *PLoS One, 6*(4), e18224. doi:10.1371/journal.pone.0018224.

Zak, P. J. (2015). Why Inspiring Stories Make Us React: The Neuroscience of Narrative. *Cerebrum, 2015*(Jan–Feb), 2.

Index

Abilities, 79, 127, 193. *See also* Mind reading ability
Aboutness, 54, 120, 171, 172, 176, 181, 198. *See also* Content; Representation
Academic historians, 2, 7, 248
Accidents, 213
Acoustical disturbances, 169
Action potential, 125
Adaptation, 186
Adaptive behavior, 226
Adaptive falsehoods, 85
Adaptive responses, 189
Aequant, 204
Aesthetics, 225
Aethelred, King, 237
African savanna, 6, 65, 85, 185, 194–195
Agriculture, 85, 88, 215, 222, 226
Akerlof, G., 9
Alarm calls, 193
Alchemy, 202–204
Aldrich, N., 10
Algorithm, 169
All-powerful agent, 86. *See also* Deity; God
Alpine barriers, 222
Amateurish hypotheses, 122
American civil war, 26, 27
Americas, 222
Amnesia, 133, 158
and conscious experience, 101–107

Anger, 243–244, 247
Animal species, 220
Anjou, Count of, 225
Anthropomorphism, 185
Apes, 192
Aplysia californica, 126. *See also* Sea slug
Apparent exceptions, 222
Appetite for stores, 10
Arabs, 246
Archduke Austrian, 95
Archival scholars, 29. *See also* Academic historians
Ardennes forest, 15–21, 168
Argument from design, 84–85
Aristotle, 37, 84, 186, 216
Armenians, 246
Armored units, 20
Arms races, 234
Art, 215, 244
Artic, 191
Artificial intelligence versus computer simulation, 168, 172
Arts, 198
Asia, 222
Asimov, I., 178
Associative learning, 149, 154
Astronomers, 203
Athenians, 90
Atomic theory, 176
Auditory consciousness, 180
Australia, 221

Austria-Hungary, 95, 106, 110, 112, 139, 211
Austro-Prussian war, 232
Autism spectrum, 55–57
Axelrod, R. 69
Azizi, H., 156

Baars, B., 107
Babies, 50
Backward replay, 151, 153
Bailey, C., et al, 130
Baldwin effect, 74–76, 84, 196
Balkans, 95, 97, 118
Banishing theory of mind from psychology, 190
Baron-Cohen, S., 52, 57
Barrier cells, 144, 145, 150, 186
Bats, 144, 179
  brain, 180, 181
Bateson, M., et al., 87
Battle of the Bulge, 17, 24
Beard, M., 8
Behavioral ecology, 223
Behavioral economics, 85, 92
Behavioral modernity, puzzle of, 194
Behaviorism, 82, 154–155, 178, 231
Belgium, 15
Belief box, 36, 37, 40, 42, 46, 114, 160, 163
Belief expression, 157
Belief, 9
  as neural circuit, 119
  rat's, 156
Belief/desire pairing, 37, 40, 43, 80, 97, 98, 110, 111, 114, 159, 174, 181–182, 201, 202, 213, 231, 233. See also Belief box; Desire box
Belief/desire synonyms, 89
Ben-Yakov, B., 148, 157
Berlin, 42
Bible, 217
Bicycle riding, 79
Binocular vision, 81

Biography, 4, 11,139, 159, 249
Biology, 33, 217, 223
Biotechnology, 84
Birth canal, 66
Bismarck, O., 15
Bison, 221
Blade of grass, 230. See also Kant, I.
"Blank check" to Austrians, Wilhelm's, 95, 97, 101, 110, 202, 230, 234
Blindness, of variation, 189
Bloom, P., et al., 79
Blud und Boden, 248
Bluntness of theory of mind as tool, 215
Boeing 747, 41
Bohemia, 214
Bonobos, 65
Bottleneck, 66. See Population bottleneck
Bottom-up strategy, 116
Boundary cells, 158
Bounded rationality, 92
Boxology, of theory of mind, 36, 37, 115, 117, 161,164, 165,166,178, 181
  diagram, 37, 101
Brain, 49, 67, 100, 155, 176
  ape's, 76
  injuries, 116
  malfunctions, 157
  reading, 121
  regions, 60
  size, 66,68
Brandenburg gate, 43
Brandler, W. et al., 57
Brazil, 193
Breast feeding, 67
Brinton, C., 111, 112, 141
Britain, 17, 19, 35, 210, 211, 229–230, 236
British bankers, 38, 42
Bronze age, 90
Brute force calculation, 166
Bryson, B., 4, 10
Bubonic plague, 225

Buckingham palace, 41
Bulow, B, 95
Burkhard, J., et al. 76
Business schools, 89

CA1 cells, 150. *See also* Hippocampus
CA3 cells, 150. *See also* Hippocampus
Cage floor, 134
Calculation, 200
Camouflage, 191
Canines, 73. *See also* Dogs
Capitalism, 219
Cargo, 219
Carruthers, P., 108
Castlereagh, R., 211, 229, 231, 234
Castration anxiety, 226
Causality, 11–12, 34, 55, 61, 84, 85, 93,
    98–100, 105, 154, 177, 182, 189,
    205, 213, 215, 220, 247
Cells, 183, 235
  anatomy, 179
  physiology, 131, 183
Central bank, 10
Cerebral cortex, 113. *See also* Neocortex;
  Motor cortex
Chamberlain, N., 214
Charles XV, 111
Cheating, 87
Chemists, 176, 206
Chess, 166–168
Chicken and egg phenomenon, 81
Child rearing, 227. *See also* Cooperative
  child rearing
Childhood, 33, 49
Children, 66, 67
Chile, 213
Chimpanzees, 65, 75, 77, 193
China, 224, 225–229, 239
Chiropterologists, 179–180
Choice point, 153
Chomsky, N., 63, 192–193
Christians, 246
Christopher Columbus, 224–225

Chromosomes, 176
Chronologies of events, factual, 2, 3,
    233, 245, 247
Churchill, W., 8, 21, 22, 23, 177
Churchland, Paul, 239
Circular box, 142
Circular reasoning, 113
Class system, 226
Classical conditioning, 126, 127, 154, 160
Clinical psychology, 132
Cloud, D., 195
Coevolution, 79, 184, 194
Cognitive activities, complex, 34
Cognitive dissonance, 218
Cognitive linguistics, 192
Cognitive maps, 192. *See* also Maps
Cognitive neuroscience, 183
Cognitive scientists, 55
Cognitive skills, 52
Cognitive social psychology, 49, 52, 92,
    243
Collaboration, 185, 186, 192, 197
Collaborative child rearing, 68, 76, 80
Collective memory, 247
Collective responsibility, 249
Combustion, 203, 205
Cometary impact, 224, 236
Commercial opportunism, 24
Common sense, 18
Competing theories, 176–177
Compound axe, 71, 72. *See also* Tools
Computer engineers, 166, 173
Computer/brain analogy, 164–174
"Computing" location in hippocampus,
    144, 158
Concert of Europe, 230, 233, 234
Conditioning, 155. *See also* Classical
    conditioning; Operant conditioning
Confabulations, 201
Conflict resolution, 196
Congress of Vienna, 112, 209–214, 229,
    230, 234–235,236, 246
Connect the dots, 12

Consciousness, 40, 46, 122, 128, 197
of anger, 244
awareness, 93, 94, 101, 173,
chapter 6
deliberation, 157
experience, 113, 180, 218
feelings, 100
memories, 158–160
representation in, 105 107
thoughts, 103
Consolidation, 130, 151
Consorts, 227
Conspicuous consumption, 226
Conspiracy theorists, 86, 200. *See also*
Hyperactive conspiracy detectors
Content, 54, 114, 118–139, 161,175,
178–181, 182, 183, 194–195, 197,
198, 218, 231. *See also* Aboutness;
Representation
of beliefs and desires, 43–44
and conscious experience, 101–107
intentional, 42–45
intrinsic, 171
original, 173
Contextualizing events, 23
Continental drift, 224
Conventions, 102
Cooper, D., 111, 112, 141
Cooperation, 67, 69, 77, 87, 194, 196,
197, 200, 243
hunting, 68
Coordination, 196, 197, 200
hunting, 78, 80
Cop-out, 240
Copernicus, N., 204, 205
Coping with the future, 25, 249–250
Copy errors, genetic, 56
Core application of theory of mind, 199
Cost-benefit analysis, 86
Creationism, 217
Creative artists, 240, 246
Creativity, 153, 166
Crick, F., 176, 182

Crimean war, 232
Croatia, 246
Crude instrument, theory of mind as, 92
Cultural anthropology, 225
Cultural constructions, 88
Cultural evolution, 6, 228, 235. *See also*
Darwinian cultural evolution
selection, 188
Cultural explanations, 226
Culture, 88, 177, 198, 202, 216,248
institutions of, 184
myths of, 177
Curiosity, 232
reduction, 86, 237
satisfaction, 210, 201, 218
Cut the cake game, 69, 71
Czechoslovakia, 214

Dangers of history, 247–249
Danish-German war, 232
Darwin, C., 34, 84, 87, 186, 189, 220,
240
Darwin's banishment of purpose from
biology, 248
Darwinian analysis, of Congress of
Vienna, 233–235
Darwinian approach
to cultural evolution, 87, 221, 224
to history, 239
to human affairs, 229
Darwinian process, 166, 194, 206, 235
Darwinian theory, theory of mind
reconciled, 238
Data structures, 170
Dawkins, R., 252
Dawn of written history, 92
Dead reckoning, 145
Decision making, 141, 150, 153–154,
161, 175, 177
Decision science, 85
Declarative beliefs, 157, 158, 159
Declarative information, 148
Declarative memories, 122, 156

Deep Blue, 166, 167
Default mode network, 61
Deferent, Ptolemaic, 204, 205
Dehaene, S., et al., 109
Daladier, E., 214
Demythologizing, 186
Denisovans, 65
Dennett, D., 52, 90, 197, 253, 255
Depression, 108
  economic, 25
Derrida, J., 173–174
Descartes, R., 177
Design argument, 186
"Design problems," 86, 87, 185, 190,
  192, 195,196, 215
Desire box, 36, 40, 42, 46, 114, 160,
  163. *See also* Boxology
Desktop computer, 164, 174
Detection of purpose, child's, 79
Detroit, 102
Development, 64, 189
  disorder of, 56
  of theory of mind, 76
Diagnostic payoff, 218
Diamond, J., 219–224, 232, 239
Deity, 185
Dinosaurs, 65
Diplomacy, 209. *See also* Congress of
  Vienna
Direction of fit, 54, 161, 166, 178
Disagreement, 196
Dispossession, 5
Dittrich, L., 123
Divine powers, 238, 239
Divine processes, 238
Division of labor, 68, 79, 215, 248
DNA, 56, 62, 176. *See also* Genes
"dogs dogs dog dog dogs," 106
Dogs, 77, 126, 154. *See also* Canines
Dolphins, 73
Domestication, 22
  of animals, 221
  of plants, 219

Dorsolateral prefrontal cortex, 59
Downloading, 166
Drama, 244
Drift, 24
Drowning, 128
DSM–5, 55
Dualism, 117, 179

Earth, 216
Earthquakes, 25, 224
Echolocation, 180
Economic choice, 91. *See also* Rational
  choice
Economics, 9, 85, 91
Edgeworth, F., 91
Egan, G., 257
Eichenbaum, H., 157. *See also* Manns
  and Eichenbaum
Einstein, A., 34, 205
Eisenhower, J., 24
Elections, 25
Electrochemical bursts, 156
  firing of, 144
  signaling, 149
Electrochemistry, 138, 179
Electrodes, 134
Electroencephalography, 117
Electromagnetic charges, 170
Electronic circuits, 168, 174
Elephants, 73, 77, 221
Ellipses, 204
Emergence, persistence, disappearance
  of a practice, 225
Emotions, 110, 193, 247, 248
  impact of history on, 243
Empathy, 248
Emperor, Chinese, 227
Empirical science, 240
Encyclopedia Britannica, 251
Ends/means systems, 84, 199–201
Engerman, S., 259
English Civil War, 27
English language, 132, 133, 170

Enigma code, 20
Entertainment, 216, 250
Entorhinal cortex, 122, 134, 135, 148, 175, 189
Environmental appropriateness, 189
Environmental filtration, 230
Epicycles, 204–205, 212, 321,247
Episodic information, 148
Eschatologica, 248
Essentialist taxonomies, 84
Ethics, 225
Eurasia, 219, 221
Europe, 222
Europeans, 219
Everett, D., 193
Evolution by natural selection, Darwinian, 188–192
Evolutionary anthropology, 64, 93, 192, 193, 223
Evolutionary biology, 217
Evolutionary game theory, 69, 225–226, 234
Evolutionary pedigree, 64, 82, 85, 166
Exogenous shocks, 224, 225, 234
Expectations, 91
Experimental economics, 91
Experimenters, 133, 142, 144, 145,148, 150
  mapping, 136
  perspective, 134
  reading, 136
  representation, 136
Experiments, 148
"Explanatory gap," 179–181
Explicit memories, 122, 123, 124, 125, 130, 131, 156
Extinction threat, 67
Eye-tracking, 50–51, 75, 79, 84

Facts of the matter, 97, 104, 112, 213
Falk, D., et al., 67
Fallibility of historical explanations, 113

False belief test, 52–53, 56, 60, 75, 78, 85, 87
Faulkner, W., 9
Feedback, 190
Feelings, 178. *See also* Anger
Fehr, E., 243
Ferdinand and Isabella, 225, 239
Fertile crescent, 219–220
  agriculture in, 221
Fetal development, 206
Feudalism, 85
Fiction, 175, 245. *See also* Historical fiction
Fine tuning of theory, 200–201
Fire making, 78
Five-pointed star marking national capital, 121–122
Flowchart, 171
  computer, 164, 165
  human (*see* Boxology)
Fluid intelligence, 61
fMRI, 57, 108, 109, 117,159, 244
Fodor, J. 88, 253
Fogel, R. 259
Folk biology, 83, 84, 92
Folk physics, 83, 92
Folk psychology, 35, 93
Food chain, 81, 192, 196, 199, 200. *See also* African savanna
Food odor, 149
  pellet, 149
Foot binding, Chinese, 225–229
Ford, G., 112
Foreign policy, 209, 213, 214
Forward replay, 151
Fox, S. 145, 147
France, 19, 33, 42, 43, 212, 229–230, 233
Franco-Prussian war, 15, 232
Frank, L., 139,142, 151, 154, 257
Frederick William III, 212
Free agents, 177
Free riding, 87

Free will, 215
French Revolution, 111
Freud, S., 117, 185, 226
Frozen accidents, 237
Full-blown theory of mind, 198
Function, biological, 188–192
Fundamentalists, 217
Future, 23
    foreseen, 7

Galileo, 205, 207, 215
Gambler's fallacy, 69, 70, 85, 91, 194,
    228, 229
Gaze tracking, 200
Gazelle, 199
Genes, 49, 62, 115, 176, 235
    knock outs, 127
    pool, 66
    silencing, 127
Genetic code, 156
Genetics, 84, 239
Genocide, 246
Geographical barriers, 221
Geographical determinism, 223
Geography, 143, 155
Geological epochs, 206
Geopolitical analysis, 223
Germany, 12, 47, 110
    army, 15–24
    language, 5
    occupation of France, 27
Germs, 219
Gesturing, 200
Gibbon, E., 5, 7, 8
Gigrenzer, G., 85, 92
Gilgamesh, 89
Giraffe, 191
Glass, C., 10
Global workspace theory of
    consciousness, 107, 109, 159
God, 238, 240. *See also* Deity
Goldman, A., 253
"Good war," 28

Gopnik, A., 86
Gorillas, 65, 75, 77
Government, 217
Grammars, 193
Grand master, chess, 166
Grand quartier-general, French, 15, 16
Grandin, Temple, 62
Great apes, 80. *See also* Apes; Primates
"Greatest Generation," 28
Gredeback, G., et al., 50–53, 84
Greeks, 246
Greene, B., 252
Grey, E., 15
Grice, P., 197
Grid cells, 134–138, 141–154, 158, 186,
    189
Griffiths, D. W., 245
Gross anatomy, 131
Grosser Generalstab, German, 16
Grunting, 200
Gulag Archipelago, 109, 110, 241–244
Gweon, H., et al., 58

H.M. (patient), 122, 124, 134
Haidt, J., et al., 244
Hand axe, 41. *See also* Tools
"Hard problem," 179–181
Hardware, 174
Hardwiring, 1, 35, 50, 57, 63, 73, 74,
    164, 196
    of theory of mind, chapter 4
Harvard University, 111
Haugeland, J., 178
Hauser, M., et al., 63, 192
Hawking, S., 4, 10
Head direction cells, 144, 145, 150, 158,
    186
Hebbian conditioning, 232
Hegel, G., 249
Henke, K., 159
Henrich, J., et al., 69, 70, 243
Henry II, 236, 237
Heredity, 176

Hexagonal regions, 134–137

Higher level of organization, 161.
    See also Levels of organization

Hippocampus, 122, 124, 128, 133,
    134, 135, 144–154, 175, 186, 189,
    199

Historical explanation, 21, 33
    rightness of, 29
    theory-free, 33–35

Historical facts, 11. *See also* Facts of the
    matter; Chronologies

Historical fiction, 177, 213, 26, 243
    versus history, 201

Historical ignorance, 214

Historical inevitability, 235

Historical knowledge, 23, 24

Historical memory, 246

Historical narrative, 21
    harmful, 6, 246–250

Historical novels, 110. *See also* Fiction

Historical teleology, 233

Historiography, 97

History
    as entertainment, 12, 250
    of science, 7
    without theory of mind, chapter 11,
    225

"History," ambiguous word in English,
    4–5

Hitler, A., 5, 17, 31–33, 46–47, 213,
    244–245

Holocene, 41, 85, 215, 222

Homer, 6, 89, 90, 92, 98, 177

Hominins, 66, chapter 5, 192, 193, 198,
    243, 248

Homo floresiensis, 65

Homo sapiens, 65, 66, 192, 194, 198,
    200, 232, 248

Homology, rat/human brain, 142, 163.
    See also Human/rat brain comparison

Homosexuality, 55

Horne, A., 24

Horoscopes, 216

Hortensius, R., et al., 244

How brains store beliefs, 131, 132–139,
    chapter 8

Hrdy, S., 64, 68, 76, 255

Hubel, D, and Wiesel, T., 63

Hublin, J.-J., et al., 65, 194

Hull, C., 138

Human agency, 215
    behavior, 161
    brain, 124, 168, 183

Human brain/rat brain comparison,
    132, 141–143, 157,163. *See also*
    Homology

Human contestant, 169

Human nature, 1

Human prehistory, 6

Human psychology, 186, 218

Humanities, 240

Hume, D., 34, 98, 99

Hunter-gatherers, 41, 215, 221, 248

Hutus, 246

Huxley, T., 220

Hyperactive agency detectors, 86, 87,
    185, 215, 249

Illusion of purpose, 230, 231, 232

Imitation, 73, 79, 81, 195. *See also*
    Teaching/learning

Immunity, 221, 223

Imperfections, evolutionary, 83

Imperial edicts, 226

Implementation of programs, 166

Implicit memory, 127
    as ability, 130

Improving a scientific theory, 88–90

"In principle" solutions, 114

Indispensability of theory of mind, 81,
    184

Industrial revolution, 85

Industrialization, 229

Infants, 35, 54
    development of, 79

Infinite regress, 174

Information, 134, 170, 171, 173
  storage, 148, 150–152, 172
Innateness, 34, 35, 49, 50, 54, 63, 74.
  76. *See also* Nearly innate
Inputs/outputs, 172, 175, 189
  circuits in brain, 125
Insect prey, 180
Institution design, 234
Instructions, to experimental subjects,
  218
Instrumental learning, 154. *See also*
  Operant conditioning
Intellectual catastrophe, 88
Intelligibility of behavior, 39
Intentionality, shared. *See* Shred
  intentionality
Intentions, 86
  action, 153
International relations, 213
Interpretation, 102, 119–122, 138,
  156,172, 173, 174, 191, 192,
  194–195
Introspection, 93, 98, 102, 194, 239,
  240
  reliability of, 100
Inverted yellow triangles, 102–104.
  See also Street signs
Iron atoms, 188
Irredentism, 246
Isaacson, W., 4, 8
Israel, 5, 246
Iterative process, natural selection as,
  188

Jack, A., et al., 62
Japan, 33, 46
Jealousy, 247
Jeopardy!, 163, 168–172
Jews, 6
Jim Crow, 26–27
Jobs, S., 4, 8
Jodl, A., 31
John, King, 237

Julius Caesar, 41
Jupiter, 205
Jury system, 236–237

Kahneman, D., 85, 91, 92
Kalbe, E., et al., 59
Kandel, E., 125–130, 141, 155, 175, 181,
  190, 206, 217, 232
Kant, I., 230
Kasparov, G., 166
Kay, K., et al., 151
Keitel, W., 31
Keller, H., 55
Kennedy, J. F., 95
Kepler, J., 205
Kershaw, I., 31–33, 46
KGB, 241
Kin
  groups, 235
  selection, 71
Kissinger, H., 112, 209–214, 229, 231–
  233, 235
Kitamura, T., 130, 139
Kitcher, Patricia, 117
Kittens, 63
Know-how, 73. *See also* Abilities
Kosovars, 246
Krupenye, C., et al., 75
Ku Klux Klan, 245

Literature, 215, 249
Literary theory, 34
Local politics, 224
Localization, of beliefs and desires, 58,
  59, 60,114, 117, 118–139
  of theory of mind, 57–62
Locke, J., 99
Logical incompatibility of theories,
  181
London Review of Books, 3
Long childhood dependency, 195
Long-term memory, 128
Longue durée, 23, 90

Lorenz, K., 63
Louis Philippe, 111
Love
  of history, 1
  of narratives, 49
  of stories, 214, 216, 250
Lowenstein, R, 4, 9
LTP (long-term potentiation), 127, 128,
    130, 131
Lurz, R., 73
Luxembourg, 15

Machine code, 166
Machine gun, 19
Mackie, G., 226–229, 232
Maginot line, 20, 22
Magna Carta, 236, 237
Magnetic field detectors, 188
Mainframe computer, 174
Males, 227
Malle, B., 78
Mammals, 75, 149, 192
  nervous system of, 189
Manifest destiny, 248
Manns and Eichenbaum, 134, 139, 142,
    149, 189
Maps, 121–122, 150, 191. *See also*
    Cognitive maps; Tolman, E.
  key, 122, 138
  legend, 122, 138
  metaphor, 155
Mariners, 216
Marketing departments, 89
Marmosets, 77
Master narratives, 249
Mathematics, 33, 61, 171
Mathematical functions, 169
Matters of convenience, 190
Maximizing fitness, 83
McPherson, J., 27, 28–29
Meaning, 45, 160, 191, 197
  of cultural practice, 225
  in history, 248

of history, 26–29
of the past, 26
Means/ends behavior, 74, 226, 227.
    See also Predator/prey tracking
Measuring devices, 203
Medial entorhinal cortex, 141–154, 186.
    See also Entorhinal cortex
Memory, 122, 199
  storage, 131
  for stories, 214
Mendel, G., 115, 176, 182
Mendel's laws, 176
Mental illness, 49, 115
  images, 104, 197
"Mentalizing," 35
Merry-go-round we can't dismount,
    218
Metal ores, 206
Metaphor, 212, 234
Metternich, K, 111, 159, 210, 211, 214,
    229, 231, 234
Microchips, 171
  circuits, 172
  electrodes, 207
Middle East, 9
Military doctrine, 19
Military history, Western Europe,
    1870–1945, 15–21
"Mill" argument of Leibniz, 178–179
Millikan, R., 258
Mind, distinct from brain, 113. *See also*
    Dualism; Explanatory gap; Hard
    problem
Mind reading, 73, 76, 77, 114, 184, 192,
    195, 196, 200, 215, 244
  compared to theory of mind, 79, 215,
    248
Mind-body problem, 177–181
Mind's ear, 106
Mindlessness, 190
Mirror-image writing, 158
Missionaries, 229
Mistreatment, 246

Models, 25
  systems, 131
Modernization, Chinese, 228
Modules, 57, 58, 60, 117, 145
Molecular biology, 182
Molecular neurobiology, 125
Molecular genetics, 179
Moltke, H., 15
Money, quantity theory of, 10
Moral authority of narrative history,
  246–249
Morality, 185
  as motivation, 27
  responsibility, 215
Moravia, 214
Morris water maze, 128
Morrow, J., 92
Moser, E., 125, 131, 133, 141–145, 174,
  175, 181, 206, 217, 232
Moser, M-B., 131, 133, 141–145, 148, 174,
  175, 181, 192, 206, 215, 217, 232
Motives, 86
Motor cortex, 244
Multiplay-multiplayer games, 70
Music, 244
Munich, 47
Muslims, 246
Mutation, 235

Nagel, T., 179
Napoleon I, 101, 111, 112, 139, 159,
  170, 171, 172, 173, 186, 209
Napoleon III, 15
Narrative explanation, 13, 175, 177,
  184. *See also* Stories with plots;
  Narrative history
Narrative fiction, 244. *See also* Historical
  fiction
Narrative histories, explanatory, 184
Narrative history, 3, 5, 13,111, 113, 177,
  181, 184, 185, 218, 229, 238
  always wrong, 215
  as stories, 3, 244

structure, 245 (*see also* Plot; Stories
  with plot)
National attitudes, 212
National narratives, 248–249
National rights, 249
National strategies, 213
Nationalism, 245
Natura non facit saltum, 130
Natural experiments, 240
Natural selection, 71, 73, 75, 82, 200,
  215. *See also* Darwinian theory
  overshoots, 86
Nazis, 249
Neander, K., 258
Neanderthals, 65
Near innateness of theory of mind, 64,
  74, 80, 81, 82, 83, 184
Necker cube, 65
Negative mass, 203
Nelson, H., 12
Neoclassical economics, 91
Neocortex, 130, 149, 151,153, 159,160,
  175, 199
Nested beliefs, 52, 53, 72
Nested sentences, 196
Neuronal firing rate, 145, 147, 151
Neural anatomy, 157
Neural circuitry, 113, 122, 138–139, 149,
  153, 155–56, 161, 163, 168, 173,
  174, 182, 198, 205, 206, 232, 244
  coding, 149
  firing, 142
  mechanisms, 115
  networks, 118
Neuroanatomy, 183
Neurogenomics, 127
Neuronal burst timing, 147, 151
Neurons, 117, 188
Neuroscience, 49, 100, 110, 114, 115,
  122, 149, 153, 163, 175, 176,178,
  180, 182, 184, 191, 192, 198, 206,
  214, 217, 239
  banishing purpose, 206–207

Neurotransmitters, 125

New Guinea, 219

New York Review of Books, 3

New York Times Book Review, 3

Newton, I., 34, 84, 186, 190, 205, 206, 215, 230

Newtonian mechanics, 216

Nichols, S., 36

Nineteenth century, 245, 249

Nixon, R., 112, 232

Nobel lecture, 138, 144

Nobel Prize, 11, 85, 115, 124,125, 126, 131, 149, 173, 183

Non-REM sleep, 151

Nonconscious thought, 180

Nonfiction best sellers, 11

Nonmaterial processes, 117

No purposes in physical world, 206

North Atlantic, 222

Norwegian language, 133

Novel strategies, 234

Noxious stimuli location, cells, 155

Nuclear extinction, 209

Nucleic acids, 183

Nucleotides, 26, 156

Object location cells, 148, 155

O'Brien, P., 12

Occurrent versus nonoccurrent beliefs, desires, 41

Oceangoing ships, 222

Oceania, 221

O'Connell, G., et al., 60

October Revolution, 241

Odor cells, 155

Oesch, N., et al., 53

Offense a l'outrance, 20

Ohman, A., et al., 63, 243

O'Keefe, J., 125, 131, 133, 136, 137–138, 141–145, 174, 175, 181, 190, 192, 206, 207, 217, 232

Old Testament, 6

Olympiads, 205

Omer, D., et al., 144

Operant conditioning, 154, 160, 190, 232. *See also* Learning

Opposable thumbs, 81

Ordinal utility, 33

Organelles, 188

Organs, 183

Oscillations, 145, 150. *See also* Theta waves

Overshooting, 55, 87, 185, 200

Ovulation, 67

Oxygen, 206

Oxytocin, 86, 109

Pacific, 222

Painting, 118, 119

Pakistan, 213

Paleoarcheology, 192, 194

Palestine, 5

Paley, W., 85

Pan balance, 207

Paris, 42, 43, 103
   direction sign, 120

Parliament, 237

Participant narratives, 225

Path-integration, 158. *See also* "Computing" location in hippocampus

Patriarchal control, 226

Pavlov, I., 126, 154, 206

Payoff, 228

Peaceful conflict resolution, 237

Pearl Harbor, 31, 46

Pleistocene, 41, 196

Peoples Republic of China, 213. *See also* China

Peoples, 248

Personification, of nations, 212

Phantom limb syndrome, 99

Pharmacology, 131, 198

Phase shift of neuronal firings, 147–148

Phenomenology of consciousness, 105–107

Philosophy, 82, 233, 249
  of biology, 217
  of history, 220
  of language, 192, 194
  of mind, 181, 161, 178
  of psychology, 35
Phlogiston, 206, 231, 238, 247
  theory, 202, 203
Photographers, 119
Physics, 216
Pinker, S., 252
Piraha people, 193
Pixels, 170, 173
Pizarro, F., 222
Place cells, 134–138, 141–154, 160, 186,
  189
Place field, 147
"Planetai" (Greek "wanderer"), 203
Platitudes about mind, 36, 45
Pleasurable emotion, 109, 250
  derived from stories, 216, 218
Pleistocene, 79, 88, 185, 215
Plowing, 221
Point mutations, 224, 225
Poland, 15
Polar bear, 191, 192
Political equilibrium, 210, 211
Political science, 91
Popular historians, 7, 9
Population bottleneck, 71, 72, 76,
  78, 81
Populations, 235
  growth, 222
Postcard, 118
Posterior medial prefrontal cortex, 60
Potentiation, 127. *See also* LTP; STP
Powell, A., et al., 68
Practical syllogism, 37
  reasoning, 38, 39
Predator/prey, 74, 75, 192
  tracking, 195, 199
Prediction of immediate future
  behavior, 87

Prediction, 22, 23, 25, 26, 29, 42, 67, 75,
  82, 85, 89, 92, 158, 181, 184, 196,
  204, 205, 214, 216, 238, 239
  historical, weakness of, 21–24
  long-range, 88
  of normal behavior, 198
Preferences, 91
Prefrontal cortex, 153–154
Preplay, 148
Primates, 73, 75, 192, 227, 248
Primatologists, 75, 193
Printed letters, 119
Prinz, J., 107
Prisoner's dilemma, 69, 71
Problem/opportunity, evolutionary, 67
Program, computer, 164–174, 175
Projections, neural, 142, 148
Propositional attitudes, 101
Protein, 67, 68, 183
Proto-language, 79, 80, 194
Prussia, 211, 212
Psychoanalysis, 116
Psychological impossibility, 184
Psychologists, 35
Psychology, 10, 35, 231
Psychopathology, 198
Psychopharmacology, 131
Ptolemaic theory, 202, 203, 215, 216, 233
Public intellectuals, 217
Punch and Judy show, 52–53
Pupil dilation, 51
Purpose, 84
  banished from biology, 206
  banished from nature, 230
  banished from psychology, 206

Quantitative measurability, 91
Quick and dirty solutions, 83, 87, 185,
  195, 215
Quine, W., 255

Races, 248
Radio traffic, 20

Random variation, 188
Rat, 183
  anatomy, 190
  experiences, 158
  head turning, 142
  and human hippocampus compared,
    156 (*see also* Homology)
  as model, 131–137
Rational choice, economic, 229
Rational decision making, 85
Reading rat beliefs, 133
Real prices, 33
Realism in international relations, 210
Realpolitik, 210
Reconciling theory of mind and
    neuroscience, 191
Recursion, 193–140
Red octagons (stop signs), 45, 102–104,
    119, 120, 171
Redefinition of representation, 217
Reductionism, 161
Reference, 197
Reference works, 172
Reflexiveness, 22
Reformation, 8
Regress, 121
Reinforcers, 154–155. *See also* Operant
  conditioning
Religion, 86, 215
  Abrahamic, 217
Remapping, 156
Replacement of theories, 216
Replay, 148, 151
Representation, 15, 44, 45,118–139,
    145, 149, 155, 172, 174, 176, 178,
    184, 191, 192, 194–195, 197, 198,
    201, 217
  and conscious experience, 101–107
Representational art, 118
Research program, 180, 203
  in history, 222
  of neuroscience, 114, 115
Responsibility, 249

Retribution, 247
Revanchism, 246
Revisionism, historical, 8–9
Revolutions, 25
Ribbentrop, J., 31
Rieff, D., 246–247
Right temporal medial junction, 60
Roman Catholic Church, 111
Roman empire, 5, 7, 8
Rough correctness of theory of mind,
    82, 175–176. *See also* Useful tool
Roumis,D., et al., 151, 156
Rube Goldberg device, 179
Rubin, R., 4
Rules of thumb, 85
Russia, 15, 101, 112, 139, 211, 12,
    229–230. *See also* Tsar Alexander

Salts, 188
Santayana, G., 21, 22, 23, 246
Sarajevo, 95
Savanna, 65, 66, 68, 85. *See also* African
    savanna; Food chain
Saxe, D., 58, 254
Scavenging, 66, 67, 69
Schelling, T., 9
Schizophrenia, 108
Schlender, B., et al., 4
Schlieffen, A., 16, 20
Schuwerk, T., et al., 58, 60
Science writers, 10
Science, 7, 25, 219
  not stories, 4
Scientific dead ends, 202
Scientific discovery, 22
Scientific explanation, 11
Scientific revolution, 216
Scientific study of history, 240
Scientific theories, 8
Sea slug, 126–128, 168, 183. See also
    Aplysia californica
Searle, J., 255, 258
Sedan, France, 15–21

Selective environments, 235

Semiconductors, 164

Sensory experience, 108
  modalities, 149

Serbia, 97

Sex, 109

Shame, 247

Shared intentionality, 51, 72, 77, 78, 80

Sharp wave ripples (SWRs), 151–154,
  158, 160, 175
  compressed, 153

Sherlock Holmes, 168

Short-term implicit memory, 127.
  See also STP

Shropshire, 230

Signal systems, 80, 195

Signs versus symbols, 191

Silent speech, 101, 104, 106, 159, 194,
  197. *See also* Subvocal speech

Similarity of rat and human brain, 104,
  132, 141–143, 150, 157. *See also*
  Homology; Human brain/rat brain
  comparison

Single cells, 188

Skinner, B. F., 154, 206

Skyrms, B., 70, 194

Slavery, 10, 26, 27

Sleep, 151

Slime molds, 80, 194

Smell circuit, 148

"Smoking gun," 182

Social environment, 227

Social institutions, 177, 184, 198

Social psychologists, 11, 56, 87. *See also*
  Cognitive social psychologists

Socioeconomic hierarchy, 227

Software, 164, 174, 175

Solzhenitsyn, A, 110, 241–244

Somatic genes, 56, 125

Sound waves, 169

Southey, R., 12

Soviet Union, 31, 232, 241–244

Spain, 10

Spartans, 90

Speculation, 200

Speed cells, 144, 145, 150, 158, 186

Spengler, O., 220

Spider/snake aversion, innateness of,
  63, 243

Spoken language, 197, 198

Square box, 142

"Stab in the back," 5

Stability, political, 210

Stable equilibrium, 234

Stag hunt, 70

Stalin, J., 241, 244

Stanford Encyclopedia of Philosophy, 251

Star chamber, 237

Sterelny, K., 195, 235

Stich, S., 253, 239

Stock market crash of, 1929, 236

Stop signs, 102–103. *See also* Red
  octagons

Stories with plots, 200, 218, 220

Stories for children, history as, 250

Stowe, H. B., 110, 245

STP (short-term potentiation), 127, 130

Straights, 246

Strategic interaction, 209, 243

Street signs, 171. *See also* Stop signs; Red
  octagons

Struggle for survival, 189

Subjective experience, 99

Suboptimal local equilibrium, 228

Subvocal speech, 157, 158, 197. *See also*
  Silent speech

Successful methodologies, 239

Surgery, 124, 198

Symbol, 102–103, 173
  versus sign, 44, 191

Symbolic adornment, 194, 200

Synapse, 125, 127, 145, 190

Syntax, 171, 175

Talleyrand, C., 111, 139,141, 159, 202,
  210, 211, 214, 231, 233, 234

Tamarinds, 77
Tank, 22
Task positive network, 61
Tasty leaves, 191
Taxi cabs, of the Marne, 17
Taxonomies
  of folk biology, 83, 84
  of psychopathology, 62 (*see also*
    DSM–5)
Teaching/learning, 67, 71, 72, 73, 76,
  79, 81, 195, 196
Technological advantage, 224
Technological change, 22, 23, 67, 78,
  219, 222, 236
Teleological reasoning, 84
  banished from history, 248
Teleology, appearance of, 85
Telescope, 205, 207
Temporal lobe, 122, 124
Temporoparietal junction, 59
Testing, 132
Tetzeli, R., 4
The Gulag Archipelago, 109, 110,
  chapter 12
Theater of mind, 158
Theory improvement, 90
Theory of mind
  on African savanna, 66–84
  banished from physics, 186
  as blunt instrument, 161
  breakdowns in, 115
  introduced, 35–46 (*see also*
    Boxology)
  as mostly correct, 82, 175–176
  predictions of, 39 (*see also* Prediction)
  reconciled with science, 199
  as useful tool, 87
Theory of natural selection, 223, 224
Therapeutic payoff, 219
Theta waves, 146–147, 151, 153, 154,
  158, 160
Thorndike, E., 154
Thought experiment, 105

Thucydides, 89, 92
Time cells, 155
Time machine, 114
Times Literary Supplement, 3
Tipping point, 228, 229
Tissues, 183
Tit-for-tat strategy, 69, 71
Toleration for others, 77
Tolman, E., 138, 155
Tomasello, M., 51, 64, 72, 76, 77, 199,
  255
Tool making, 67, 71, 72, 78, 195
  complex, 194
Top-down strategy, 116
Toynbee, A., 220
Tracking prey/predictors, 73. *See also*
  Predator/prey; Means-ends
Tracking truth, 82–83
Transcranial magnetic stimulation
  (TMS), 58, 59, 64, 108, 109
Treat, reward, 150
Treating stories as knowledge, 214
Trebek, A., 169
Trial by ordeal, 236
Triggering stimuli, 64, 147
Tripartite pact, 33
Triple Alliance, 118
Triple-aboutness, 103–104. *See also*
  Content; Representation
Trojan War, 177
Tsar Alexander, 139, 159, 160, 211, 212,
  214, 229, 231, 234
Tuchman, B., 24
Tuning up functions, 190, 192
Turchin, P., 259
Turks, 246
Tutsis, 246
Twentieth century, 226, 228, 249
Twenty-first century, 246
Twins, 57
Two-year-old's physical understanding,
  compared to great apes, 77
Type II errors, 86

Ultimatum game, 70
Ultra decryption, 22
Unambiguous beliefs, 133
Uncertainty of paternity, 227
Uncle Tom's Cabin, 109, 110, 245
Unconscious cognition, 60, 193
Understanding, 2, 3, 5, 6, 25, 190, 214
Unfalsifiability, 82
United States, 17, 47, 232
   U.S. Army histories, 21
Unpredictability, 22
Upside-down triangles, 171. See also
   Street signs
Urals, 222
Urbanization, 222
Useful tool, theory of mind as, 87, 177

Values, 217
van Schaik, C., 76
Variation, random, 192
   blind, 192
Veblen, T., 226
Vengeance, 247
Ventral medial prefrontal cortex, 59,
   153–154
Videos, 75
Vikings, 236
Vindicating theory of mind, 198
Viral disease, 221
Visual imagery, 194
Volcanoes, 224

Wantage laws, 237
War, 6, 15–21, 233
Washington, G., 8
Watson (the computer), 168

Watson, J., 11, 176, 182, 206
Watson, T., 168
Watson's cylinders, 172
Weather, 25
   forecasting, 23
Wehrmacht, 17
Weinberg, G., 47
Welt Geist, 249
Western hemisphere, 221
Westernizing, 229
What theory of mind gets right, 201
White House, Washington, D.C. 216
"White Man's Burden," 248
White people, 246
Whites of eyes, 200
Wilhelm II, 95, 182, 202, 236
Wilson, W., 10
Wiring diagram, of neural circuit, 119
Wittgenstein, L., 177
Working memory, 154, 158
World history, 220
Worldviews, 217
World War One, 5, 8, 12, 19, 24, 25, 95,
   110, 181, 217, 220, 236
World War Two, 19, 24, 31, 111, 177, 214
Wright, L., 258
Written language, 200

Yield signs, 102–103. See also Street signs;
   Red octagons
Yu, J., et al., 139, 142, 154

Zak, P., et al., 86
Zebras, 221
Zek, 241
Zygote, 57